Evolutionary Catastrophes

Why did the dinosaurs and two-thirds of all living species vanish from the face of the Earth sixty-five million years ago? Throughout the history of life, a small number of catastrophic events have caused mass extinction and changed the path of evolution forever.

Two main theories have emerged to account for these dramatic events: asteroid impact and massive volcanic eruptions, both leading to nuclear-like winter. In recent years, the impact hypothesis has gained precedence, but Vincent Courtillot suggests that cataclysmic volcanic activity can be linked not only to the K–T mass extinction but also to most of the main mass extinction events in the history of the Earth. Courtillot's book explodes some of the myths surrounding one of the most controversial arguments in science. It shows among other things that the impact and volcanic scenarios may not be mutually exclusive. This story will fascinate everyone interested in the history of life and death on our planet.

VINCENT COURTILLOT is a graduate of the Paris School of Mines, Stanford University, and University of Paris. He is Professor of Geophysics at the University of Paris (Denis Diderot) and heads a research group at Institut de Physique du Globe. His work has focused on time variations of the Earth's magnetic field, plate tectonics (continental rifting and collision), magnetic reversals, and flood basalts and their possible relation to mass extinctions. He has published 150 papers in professional journals, and a book entitled *La Vie en Catastrophes* (Fayard, Paris, France, 1995). This volume is a translation and update of this book. Courtillot is past-director of graduate studies and funding of academic research of the French Ministry of National Education (1989–93), past-director of the Institut de Physique du Globe (1996–98), and past-president of the European Union of Geosciences (1995–97). He has been a consultant for the French Geological Survey (BRGM). He is a Fellow of the American Geophysical Union, Member of Academia Europaea, and Associate of the Royal Astronomical Society and he won the Silver Medal of the French Science Foundation (CNRS) in 1993. He has lectured at Stanford University, the University of California at Santa Barbara, Caltech (Fairchild Distinguished Scholar), and the University of Minnesota (Gerald Stanton Ford Lecturer) and is a senior member of Institut Universitaire de France. In June 1997, he became special advisor to the Minister of National Education, Research and Technology, in charge of higher education and research and in December 1998, the Director in charge of research for the Ministry. Vincent Courtillot is a Chevalier de l'ordre national du Mérite and Chevalier de la légion d'honneur.

For Michèle, Carine and Raphaël

Evolutionary Catastrophes

The Science of Mass Extinction

VINCENT COURTILLOT

Translated by Joe McClinton

CAMBRIDGE
UNIVERSITY PRESS

PUBLISHED BY THE PRESS SYNDICATE OF THE UNIVERSITY OF CAMBRIDGE
The Pitt Building, Trumpington Street, Cambridge, United Kingdom

CAMBRIDGE UNIVERSITY PRESS
The Edinburgh Building, Cambridge CB2 2RU, UK
40 West 20th Street, New York, NY 10011–4211, USA
477 Williamstown Road, Port Melbourne, VIC 3207, Australia
Ruiz de Alarcón 13, 28014 Madrid, Spain
Dock House, The Waterfront, Cape Town 8001, South Africa

http://www.cambridge.org

Originally published in French by Editions Fayard [1995]
English translation with revisions © Cambridge University Press 1999

First published in 1999
First paperback edition 2002

Printed in the United Kingdom at the University Press, Cambridge

Typeface Plantin (Monotype) *System* QuarkXpress®   [UPH]

A catalogue record for this book is available from the British Library

Library of Congress Cataloguing in Publication Data

Courtillot, V.
[Vie en catastrophes.   English]
Evolutionary catastrophes: the science of mass extinction / Vincent Courtillot;
translated by Joe McClinton
    p.   cm.
Includes index.
ISBN 0 521 58392 6
1. Catastrophes (Geology, Geophysics)   2. Extinction (Biology)   I. Title.
QE506. C7513   1999
576.8′4 – dc21   98–32169 CIP

ISBN 0 521 58392 6 hardback
ISBN 0 521 89118 3 paperback

Contents

Not even the most tempting probability is a protection against error; even if all the parts of a problem seem to fit together like the pieces of a jig-saw puzzle, one must reflect that what is probable is not necessarily the truth and that the truth is not always probable.

Sigmund Freud

Moses and Monotheism (1939)[1]

1 From *The Standard Edition of the Complete Psychological Works of Sigmund Freud,* Vol. XXIII, translated from the German under the general editorship of James Strachey, in collaboration with Anna Freud, London, The Hogarth Press and the Institute of Psycho-Analysis, 1964.

Preface

I would like to tell a story here, or rather a fragment of the story of the natural history of our planet and the beings that populate it. With Darwin, the evolution of species became part of our collective awareness. People more or less recall glimpsing the trilobites or dinosaurs, sea lilies or mastodons in those superb dioramas of which our mid-century museums were so proud. People know that a vague link of ancestry ties us to these fantastic animals, which belong to the 99% of all species that once lived on Earth and have now departed from it forever. Why are most of these animals no longer around us? Do paleontologists, whose profession it is to discover and describe fossil species, know the reason for these extinctions? Do they occur rarely, or often? Did they come about suddenly, or gradually and regularly over the course of geological time?

Well – both. Species disappear every year. And this has been so since the dawn of Life. But there are a small number of periods during which the extinctions of ancient species and the appearances of new ones attain an astonishing concentration within a rather brief time. What then are the causes of these profound breaks in the line of species, those very breaks that led nineteenth century science to define the great geological eras? The answer began to come to light less than two decades ago. Several times in the course of the history of our globe there occurred catastrophes, undoubtedly difficult to imagine, that caused vast slaughter and resulted in a mass extinction of living species. Though of major importance, this notion of extinction has generally been neglected by biologists. Since the early 1980s it has fallen to geologists to prove that convulsive phases of extinction have indeed occurred repeatedly over geological time – for the record has been preserved in fossils.

The model we inherited from the nineteenth century represents geological and biological processes as unfolding in gradual and regular harmony. To Lyell and Darwin it was simply the immensity of time and the incomplete record of this time preserved in rock that might at times give the impression of abrupt change. This scheme seemed to have been swept away when, in 1980, a team led by the American physicist Luis Alvarez and his geologist son Walter announced that the disappearance of the dinosaurs 65 million years ago was the result of an asteroid impact. Almost immediately, without denying the catastrophic aspect of the changes the world has witnessed since the end of the Mesozoic, another hypothesis followed: the last great mass extinction may have been initiated by extraordinary volcanic eruptions, in which a vast portion of the Deccan region of India was covered with lava.

This was the resurgence of the century-old debate between the "gradualists," for whom nothing special happened at the boundaries between geological eras, and the "catastrophists." This debate goes back to Lamarck and Cuvier in the late eighteenth century. And over it is superimposed a second controversy: if there was indeed a catastrophe, did death come from the sky, or from the bowels of the Earth?

In order to find an answer, geochemists and geophysicists journeyed to the ends of the Earth to sample and analyze the rare surviving archives of the time of the catastrophe. They investigated metals and rare minerals, iridium and shocked quartz (whose odd names will soon become familiar to the reader), isotopes, remnant magnetization in rocks – and, of course, fossils. Have all these potential sources of evidence preserved the memory of the last great crisis the Blue Planet had suffered? Would we be able to measure the age of such ancient objects and events with enough precision to distinguish between the mere seconds' duration of an impact and the millennia that an eruptive volcanic phase might last? How many other catastrophes had marked the history of Earth and changed the course of species' evolution in a jagged line? Was the end of the trilobites and stegocephalians, which accompanied the lowering of the curtain on the Paleozoic Era 250 million years ago, caused by the same forces as the end of the dinosaurs and ammonites?

The quest for answers to these questions has been a great scientific adventure. Retelling this adventure is also an occasion, as we pass through a review less austere than some scholarly manuals might impose, to describe the great discoveries in earth science in the last quarter of the twentieth century. The attraction of these discoveries is attested by the recent appearance of Paul Preuss's novel *Core*. In this new *Voyage to the Center of the Earth*, a physicist father and his geologist son are unwitting competitors in drilling through the Earth's mantle. It says something about what a thorough dusting-off geophysics has enjoyed when, duly spiced up with a dash of added greed and love interest, it can now compete with Michael Crichton's *Jurassic Park*.

We will need to adjust to a different way of perceiving the measurement of time and discover just how dynamic the inanimate world can be. Modern chaos theory finds superb illustrations here on an unwonted scale: sudden reversals of the earth's magnetic field, and the more majestic formation of those enormous instabilities known as mantle plumes.

It is, in fact, the inanimate world that caused the great fits and starts in the evolution of Life. The Moon is deeply marked by the great impacts that sculpted its surface down through its history. On the Earth, most of these impacts have been erased by erosion and the incessant drift of the continents. But have they played no role in the history of species?

In 1783, an eruption – quite a modest one, really – devastated Iceland and upset the climate of the entire Northern Hemisphere. Yet this eruption was a hundred thousand times less than the great basaltic outpourings that surged ten times across the Earth's surface over the past 300 million years. Wouldn't these have thrown the climate out of balance beyond all imagining? So, impact or volcanism: which is the answer?

Dust and darkness, noxious gases and acid rain, persistent cold followed by suffocating heat: the scenarios of these ecological catastrophes, whether their sources lie beyond the Earth or deep within it, inspired the terrible concept of the "nuclear winter". And, as has never before happened in geological time, a species – ours – is by itself able to alter the atmosphere to the same extent as the great natural disturbances, and far more rapidly. Deciphering past catastrophes may perhaps be the only way of predicting the future effects of human activity on this planet's climate.

This history is also meant to bear witness to the exciting world of scientific research, to an adventure that is both individual and

collective. The accidents, setbacks, changes of approach, and suc-
cesses that punctuate a researcher's career are not unlike those that
episodically alter the course of evolution. And so we will be transported to lovely Umbria in Italy, to the roof of the world in Tibet,
then to the Deccan Plateau in India, and the tip of the Yucatán
Peninsula in Mexico. We will seem to change subjects, goals, and
methods. We will encounter failure at times – but fortunately only
temporarily.

Scientists' quarrels are frequently sharp, sometimes unpleasant,
often fascinating, and always rife with new knowledge. They parallel the sometimes chaotic evolution of ideas. They make it possible
to understand how a hypothesis is built, why a researcher hesitates,
how long "truth" can search for evidence only to find it unexpectedly all at once and surge ahead. In the course of this narrative I
hope to help the reader share some enthusiasms and, perhaps, even
inspire a vocation. My purpose is in determined opposition to the
aim of the great Swiss mathematician Leonhardt Euler: someone
once asked him why the published demonstration of his theorems
had been so extensively rewritten that it was impossible to understand how he had conceived his ideas. He haughtily replied that the
architect never leaves his scaffolding behind.

Impact or volcanism? Or both together? The reader will certainly
not neglect to look critically at the new catastrophic models that
appear in this study. The metaphor of the puzzle that Freud evokes
in the epigraph to this book applies particularly well to the geological sciences, where the record of far-off times is so very fragmen-
tary. Karl Popper echoes him:[1] "A theory may be true though nobody
believes it, and even though we have no reason for accepting it or
for believing that it is true." As for me, I see Freud's metaphor as
a reminder that from time to time one has to know how to throw
caution to the winds: this is often the price of decisive advances.

A new conception of the erratic march of evolution is emerging
and has been well described by Stephen Jay Gould. The tree often
used to represent the genealogy of species bears little resemblance
any more to a grand old oak. Instead, it is espaliered: the first

1 In Karl R. Popper, *Conjectures and Refutations:
The Growth of Scientific Knowledge*, New York,
Basic Books, 1962.

branches emerge low down, at right angles to the trunk, only to branch again immediately and rather often, again taking the vertical. As though the gardener had gone berserk with the pruning shears, from time to time most of the branches are lopped off, even many that are perfectly healthy. Those that remain were just lucky.

In "normal" times – in other words, most of the time – the process of evolution is governed by necessity. But the role of chance, during the rare and brief moments when it strikes, is so great that one almost wonders whether it does not play the main role after all. Humans would probably not exist and our environment would be unrecognizable if the nature of certain improbable catastrophes, and the order in which they occurred, had not left an indelible mark on the living world.

I would also like to express my heartfelt thanks to those who were kind enough to be the first readers of this book and help me to improve it by their observations: José Achache, Guy Aubert, Michèle Consolo, Emmanuel Courtillot, Jean-Pierre Courtillot, Yves Gallet, Jean-Jacques Jaeger, Claude Jaupart, Marc Javoy, Jean-Paul Poirier, and Albert Tarantola. Françoise Heulin and Claude Allègre provided me with crucial advice about the overall organization. Joël Dyon provided the illustrations. The French part of the research reported in this work was financed by several universities, the Institut de physique du globe de Paris, and the Institut national des sciences de l'univers (CNRS).

Paris, Pasadena, Villers

Preface to the English edition

Four years have elapsed since the original French version of this book came out. It is my feeling that much of the research that has appeared in print during this time has further vindicated the views I held back in 1995. I would like to thank Joe McClinton for what appears to me as an excellent and faithful job in translating the French original version of this book into English. This translation has given me a number of opportunities for updates, for example on the age of the Permo–Triassic sections from China, the eruption of the Emeishan Traps, the confirmation of the presence of anomalous iridium in the Deccan Traps (in the district of Kutch), our recent work on the Ethiopian Traps, the strong link between flood basalts and continental rifting, and the further suggestion that catastrophes (whether volcanic or of some other kind) are a prerequisite for any major shift in evolution. I hope English-language readers will enjoy this unconventional account of the causes of mass extinctions and reflect on the potential of modern Earth Sciences in helping us to use the past to make the future more understandable, though perhaps not predictable.

At the end of the book a Glossary, essentially produced by Stuart Gilder, to whom I am particularly grateful, defines many of the terms used within the text.

Paris, 1999

Foreword

The dinosaurs are the most famous of all fossils. From gigantic *Diplodocus* to terrifying *Tyrannosaurus rex*, through the waystations of the pterodactyl or *Triceratops*, they have all haunted our childhood fantasies. For more than a century, these strange fossils have posed a daunting riddle for scientists. They had reigned unchallenged for 200 million years on land, in the sea, and in the air; they were superbly adapted to their environment; they never ceased to grow larger and larger; yet all at once they vanished from the face of the Earth some 65 million years ago. Why?

In 1980 the physicist Luis Alvarez and his son Walter, a geologist, proposed an answer to the riddle: a gigantic meteorite struck the Earth, plunging it into dark and cold for several years. They thus revived the old hypothesis of Georges Cuvier, which linked changes in fossil flora and fauna to natural catastrophes. Was Darwin wrong in his theory of the continuous evolution of species?

The Alvarezes' work exploded like a bombshell in the serene skies of paleontology, sparking an extraordinary degree of scientific activity focused on their hypothesis and its consequences, and rapidly pitting supporters against dissenters. After a decade of space research, was it not natural to appeal to a cosmic influence in the evolution of species? On the other hand, was it acceptable that two scientists – themselves not even paleontologists – should call into question the 'certainties' of an entire profession? The exchange of arguments was vigorous, if not always rigorous.

It is this extraordinary scientific adventure that Vincent Courtillot recounts for us. But this is not the narrative of a spectator, however committed. It is an account from one of the active, creative, and incisive participants in this adventure, a participant who defends a thesis with talent and precision, but who also accepts arguments from others, provided they can pass the muster of his implacable logic.

This book reads like a novel, and the end takes an unexpected twist – incredible yet probable, a conclusion that shatters probabilistic beliefs, the well-known refuge of those who dwell among certainties.

I will leave to the reader the pleasure of following the episodes of this saga, which will remain one of the major scientific polemics of the current turn of the century.

Professor Claude J. Allègre
French Minister of Education, Research and Technology,
Professor, University of Paris VII – Denis Diderot and
Institut de Physique du Globe de Paris

Mass extinctions

A short history of Life on Earth

The Earth had already been revolving around the Sun for nearly four billion years when Life entered a major new stage. For more than two billion years, the only life forms had been isolated cells floating in the worldwide ocean. But now these cells began to associate with one another, becoming the first multicellular organisms. This was some 700 million years ago.[1]

It would take only another 100 Ma for certain organisms to develop a skeleton: hard parts that could be preserved in rock long after the organisms died. What we know of the past forms of Life on Earth is largely based on these fossils: they have given us a far more accurate picture of the past 600 Ma than we have of the billions of years that went before.

Another 100 Ma, and the seas are now populated with fish. Yet another 100, and their descendants can lay sturdy eggs; now equipped with lungs, they grow bolder, abandon the water, and conquer the continents, as yet uninhabited. Then, 260 Ma ago comes the "invention" of warm blood, and the first proto-mammals begin to prosper. Here, at the end of the Paleozoic Era (Fig. 1.1), the abundant and varied fauna and flora bear every mark of success, both in the ocean depths and on the emergent land. Yet almost all at once, 250 Ma ago, a catastrophe causes 90% of all species to vanish forever.[2] For an entire species to disappear, every individual it comprises must die without descendants. When 90% of all species

---

1 A million years will be our 'unit of reckoning' for geological time, and we will abbreviate it as Ma.

2 Biologists have developed a hierarchical classification of living organisms based on the concept of an 'evolutionary tree.' This taxonomy recognizes seven levels, from the kingdom (of which there are five: Plants, Animals, Fungi, Protista, and Monera) to the phylum (of which there are between 20 and 30), then the class, order, family and genus, and ending with the last, indivisible unit, the species. By definition, this last groups together those individuals capable of reproducing among themselves.

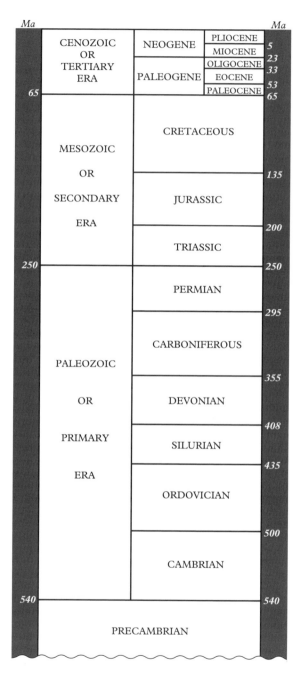

Figure 1.1
The geological time scale, with the main divisions since the Cambrian Period. Ages are given in millions of years (Ma).

die out, the populations of the remaining 10% will certainly be hard hit as well: in fact, perhaps 99% of all animals living at the end of the Paleozoic perished. This is the most extensive of all mass extinctions known today.

But not all died, and the survivors set out to reconquer the space so unexpectedly swept clear for them. This start of the Mesozoic Era is dominated by pig-sized plant-eaters called *Lystrosaurus*. They have large amphibians for company, along with other reptiles who will soon give rise to the first true mammals and the first dinosaurs. A new catastrophe, less violent than the first, arrives to decimate the last proto-mammals, the great amphibians, and (in the oceans) almost all species of ammonoids.[3]

Small, hiding in the trees and living on insects, our mammal ancestors were anything but conspicuous. You might almost say they encouraged the world to forget they were there. For this, in fact, was the real beginning of the age of dinosaurs. Recent paleontologic research has given us a whole new perspective on these beasts. Some may have been warm-blooded. The great long-necked, plant-eating sauropods, like the celebrated *Diplodocus*, gradually gave way to animals sporting horns and duckbills, grazing no longer on the treetops but on grass and bushes. Their predators were those great carnivores, colorful and agile, who for decades have delighted children and made film producers' fortunes. A few minutes of *Jurassic Park* and *The Lost World* (the movies) give a very fine view of them.

Then, 65 Ma ago, a huge catastrophe once again ravaged this world, which had seemed so perfectly adapted and balanced. This was the end of the dinosaurs and many mammals, but also of a great many other terrestrial and marine species, including the well-known ammonites and a considerable number of smaller and less familiar organisms that constituted the marine plankton. In all, two-thirds of the species then living (and possibly 80% of all individuals) were wiped out. This is the second great mass extinction.

Yet again the momentum resumes, and in less than 15 Ma we find the ancestors of most animals that still live on our Earth today.

3 The ammonites would later descend from their survivors.

As the climate turns colder, modern fauna comes into place some 30 Ma ago. The age of dinosaurs has yielded to the age of mammals, delivered at last from their chief rivals. And the Mesozoic is succeeded by the Cenozoic Era.

Extinctions and geological eras

Paleozoic, Mesozoic, Cenozoic:[4] for you, as for me, the names of the geological eras may summon up the boredom of old-fashioned junior-high science classes. Yet for all that, they still reflect the great rhythms of the evolution of species, and of great catastrophes that have shaken our globe down through its history.

It was in 1860 that John Phillips decided to define the three great geological eras on the basis of the two major biological disruptions we have just mentioned. These disruptions were discovered by George Cuvier (1769–1832), telling us something not only about this scientist's gifts but also (since they were recognized so early) of the exceptional magnitude of these catastrophes, when not only the actors in evolution but the very rules of the game itself abruptly seem to change. Species, like the living beings of which they consist, have a history: they are born, they develop, and then one day they are no more. No doubt it's hard for human beings to imagine the end of the species they belong to, or to conceive that over 99.9 % of the species that ever lived on Earth are already extinct. American paleontologist David Raup wryly observed that a planet where only one species in a thousand survives is hardly safe.

From the nature and distribution of the fossil remains he took from the rocky strata of the Paris Basin, Cuvier discovered that each stratum is characterized by an assemblage of its own typical fauna. But above all, he realized that a great many of these species no longer exist – they are extinct. Cuvier credited the Divinity for their sudden appearance and blamed their disappearance on some more earthly cause (a "terrible event," he wrote), such as a catastrophic

---

4 Geologists often prefer Greek etymology to Latin. But some, among them the French, also speak of the Primary, Secondary, and Tertiary Eras. The three Greek terms mean the ages of Ancient, Intermediate, and Recent Life. We'll use the two sets of terms interchangeably, particularly 'Cenozoic' and 'Tertiary.'

flood. It was thus that he identified the Biblical Flood as the last event preceding the modern age and the appearance of humans. According to him, none of the "agents" that Nature employs today "would have sufficed to produce its ancient works." When in 1801 his colleague Geoffroy Saint-Hilaire (1772–1844) brought back from Egypt the mummified bodies of animals identical to species still extant, Cuvier was convinced that between any two catastrophes the species remained the same and underwent no modifications.[5]

The rise of catastrophism

This catastrophism, adopted by many geologists, was in evident harmony with the predominant theology of the day and perhaps drew additional, if unconscious, support from the political turmoil amid which the "age of enlightenment" drew to a close. For instance, in 1829 Elie de Beaumont established the existence of a major episode of geological uplift in the Pyrenees, between the end of the Mesozoic and the beginning of the Cenozoic, and saw the rise of the mountains as the chief cause for the mass extinction of species between the two eras. Many naturalists back then believed that geological time had been punctuated by catastrophes, and that each event may have had a different cause.

Yet ever since the middle of the eighteenth century, another school, taking its independent and very different inspiration from Buffon (1708–88) in Paris and Hutton (1726–97) in Edinburgh, had resisted the appeal of catastrophes and attributed the magnitude of the observed phenomena to the immensity of geological time. Before Cuvier was even born, Buffon had rejected the notion of original catastrophes and estimated the Earth's age at the then-imposing figure of 75,000 years,[6] whereas the Biblical calendar set the Creation only 6000 years in the past. Twenty-five years older than Cuvier, and unaware of Hutton's works, the militant freethinker Lamarck (1744–1829) also reached the conclusion that the dynamics of

5 However, toward the end of his life, he would become persuaded that species are partly molded by their environment and may transmit some of the characteristics thus acquired to their descendants.

6 He would even propose – though without publishing it – the figure of 3 Ma, almost unimaginable in those days. See for example E. Buffetaut, *Des fossiles et des hommes*, Paris, Laffont, 1991.

geological processes are slow but inexorable. Without ever using the term evolution, he conceived the slow changing of species; unfortunately, his vision would degenerate into caricature in the hands of some of his successors. In particular, he realized that the 3000 years that separate us from Geoffroy Saint-Hilaire's Egyptian mummies are negligible in comparison with geological time. But Lamarck did not accept the idea that species might become extinct. According to him, they are gradually transformed by direct descent, or even (for those species that have apparently disappeared today) still survive in unexplored regions of the globe. His German contemporary Blumenbach (1752–1840) took a significant step in proposing that the two concepts of vanished species and distinct epochs in Nature should be combined.[7] He envisaged a long succession of periods, characterized by distinct faunas eliminated one after the other by climatically induced global catastrophes.

Where Lamarck intuited an extreme plasticity of species, Cuvier saw only absolute fixity. Able and powerful, the latter would ensure that his ideas were accepted, at least during his lifetime. It would be up to Charles Darwin to show that Cuvier's remarkable observations, which influenced him significantly, were to some extent compatible with the very theories Cuvier fought, and that Lamarck and Geoffroy Saint-Hilaire were not entirely on the wrong track. Which nevertheless did not prevent him, in his *The Voyage of the Beagle*, from taking a good many potshots at Lamarck, whom some view as the other founder of the theory of evolution.

Uniformitarianism replies

Cuvier's catastrophism was vigorously defended by Buckland in England and Agassiz (better known for his work on glaciation) in the USA. But Charles Lyell (1797–1875) took up the torch from Buffon and Hutton and carried it much further. In his *Principles of Geology*, the first edition of which appeared in 1830, he refuted the entire idea of catastrophes and postulated that all observed geological phenomena must be explicable by processes still in existence. He assumed that these processes had not varied, in either their

7 In E. Buffetaut, see note 6.

nature (a theory called uniformitarianism) or their intensity (and this theory acquired the name "substantive" uniformitarianism). Thus only the incredible length of geological time explains the magnitude of the observed phenomena: the erosion of valleys, the uplift of mountain chains, the deposition of vast sedimentary basins, movement along faults owing to cumulative seismic activity – and the mass extinction of species. As Lyell himself said, no vestige remains of the time of the beginning, and there is no prospect for an end. This world, in its state of equilibrium, held no place for evolution. A friend of Darwin, who was profoundly influenced by his work, Lyell nevertheless had the greatest difficulty rejecting the idea that species were static. Until 1860, he instead imagined a cyclic history for the Earth and the life forms inhabiting it. Darwin himself thought nothing more astonishing than these repeated extinctions, which he, in fact, explained by long periods that left no geological deposits. He discreetly discarded everything in observations that might support catastrophism and chalked up such findings to imperfections in the geological record instead.

The early nineteenth century witnessed the opposition – sometimes violent – of the catastrophist school and the uniformitarian school. Yet this theoretical quarrel did not prevent geology from growing. Quite the contrary. Lyell's views would ultimately triumph and make it possible to found a great many branches of modern scientific geology. In fact they remain deeply ingrained in the minds of most geologists, even as recent history has made us familiar with the concepts of evolution and dynamism and, unfortunately, given vigorous new life to the notion of catastrophe. Nuclear war, overpopulation, famine, desertification, the greenhouse effect, the hole in the ozone layer – so many threats, real or assumed, that frighten us and that our newspapers outdo one another in reporting – all are birds of ill omen for the agitated end of a millennium. Are humans at risk of disappearing, the victims of their own folly or of a Nature gone haywire? If, as Lyell thought, the present must be our key to understanding the past, this same past in fact harbors the keys, sometimes carefully concealed, to a better understanding of our present, and possibly to a way of safeguarding the future.

The geological time scale

To discover these keys, however, we need some kind of orientation mark. We have to measure time. Little by little, since the nineteenth century and Lyell, a history of geological time has been built up and is still being improved today. Paleontologists and stratigraphers have learned to recognize the regional or global significance of changes in fauna and flora, assess the size of these changes, and determine the continuity of their rhythm. This has allowed them to set up, and continue to refine, a time scale (Fig. 1.1), with its eras, periods, epochs, stages, and substages. The second half of the twentieth century contributed a method to measure these times absolutely; geochemists and geochronologists now know how to determine time from the radioactive decay of a number of chemical elements. More recently, in the lava of sea floors and later in exposed continental sediments, geophysicists discovered long sequences of sudden reversals in the magnetic polarity of rocks. Numerous, irregularly spaced, and very brief, these reversals made it possible, once they were identified, to establish an extraordinarily close-meshed web of correlations, and thus an effective means of determining dates (see Chapters 2 and 7).

Today we have an absolute geological time scale, particularly for the fossil-bearing ages (or in other words, approximately the last 600 Ma). In the brief description of the history of Life on Earth that we started with, we tossed about figures of hundreds of millions of years. But now we need to get more familiar with that very long unit of reckoning, a million years. Often the duration of geological time is illustrated by comparison to a single year.[8] But it seems just as illuminating to recall that our planet was formed about 4500 Ma ago; that the dinosaurs disappeared 65 Ma ago; that our ancestor (or cousin?) Lucy lived 3 Ma ago. It is also worth remembering that the last period of maximum glaciation was 20,000 years ago (0.02 Ma) and that the conflicting scenarios we are going to examine to describe what the Earth went through at the end of the Mesozoic took several Ma, according to some experts – and only a few seconds, accord-

---

8 In this case, the Mesozoic covers only two weeks of the last month of the year, from December 11 to 26, when the Cenozoic begins. The human race appears at 2 p.m. on December 31; the pyramids are built at 30 seconds to midnight.

ing to others! Between this second and the age of the Earth, the reader must blithely contemplate 17 orders of magnitude.[9]

"Normal" extinctions or mass extinctions?

Paleontologists know that apart from a few very rare "living fossils" (such as the fish called the coelacanth or that lovely tree the ginkgo), most species have a span of existence that is on the whole quite short in terms of the yardstick we have adopted: after a more or less extended period of stability, they ultimately die out. This lifespan ranges from a few hundred thousand years to several million years; the average, depending on the group, lies between 2 and 10 Ma. Within a given set of species, the probability of extinction is essentially constant over long periods (and, therefore, does not depend on how ancient the species may be) and is much greater during shorter "revolutions."[10] Extinctions during "calm" (or "normal") periods are thought to result from the normal evolution of species within a community in perpetual interaction, while revolutions are caused by a change in living conditions within the environment. The evolution of some groups of mammals during the Cenozoic, for example, is punctuated by changes in ocean currents and in climate, the causes of which must be sought partly in the famous Milankovic cycles[11] and partly in the changes in the ocean basins caused by incessant continental drift.[12]

But as we have already seen, the history of biological evolution is not limited to the humdrum course of "normal" extinctions. More rarely, there are mass extinctions in which a great many species from

9 Or 'ten to the seventeenth power,' i.e., a 1 followed by seventeen zeros, or a hundred million billion!

10 See Jean-Jacques Jaeger, *Les Mondes fossiles*, Paris, Odile Jacob, 1996.

11 The gravitational effect of the giant planets Jupiter and Saturn has a quasi-periodic influence on the angle (or 'obliquity') of the axis of rotation of the Earth and on the eccentricity (the elliptical shape) of its orbit. The Moon and Sun, for their part, exert forces that induce a precession of the Earth's axis of rotation. The periods corresponding to these three evolutions are, respectively, about 40,000 years (obliquity),

400,000 and 100,000 years (eccentricity), and 25,000 years (precession). The amount of sunshine, which varies as a function of latitude and season, is thus modulated over the same long periods. These Milankovic cycles are thought to be responsible for the changes in glaciation over the past million years (the last glacial period culminated 18,000 years ago) and also for the variations in climate recorded in far more ancient sediments.

12 See Claude Allègre, *The Behavior of the Earth*, Cambridge, MA, Harvard University Press, 1988.

most groups disappear almost simultaneously, so close together in time that chance alone cannot adequately explain it. The two most striking events of this kind mark the transition from the Paleozoic to the Mesozoic, and from the Mesozoic to the Cenozoic. To determine the age, duration, and extent of these events, David Raup and John Sepkoski have compiled the dates of appearance and disappearance of several thousand families[13] and several tens of thousands of genera of invertebrate marine organisms. The curve for the variation in number of families (Fig. 1.2, bottom) gives a quantitative view of this evolution in diversity, which we described qualitatively above. It shows a very rapid acceleration at the start of the Paleozoic, not only because of a very real diversification of species, but also because from this point on these species would be producing hard body parts. Over the next 200 Ma, diversity seems to remain constant, except for two crises, one around 440 Ma ago (the so-called Ordovician-Silurian boundary) and the other around 370 Ma ago (during the Upper Devonian Epoch). But the most dramatic event is the great catastrophe at the end of the Paleozoic (250 Ma), at the boundary between the Permian and the Triassic-whence the term Permo-Triassic crisis that we will use from now on is derived. Life, or more precisely diversity, then rapidly resumes its momentum, suffers a new crisis at the Triassic-Jurassic boundary (210 Ma), exceeds the richness it achieved during the Paleozoic and then suffers its second major crisis – which, as we have seen, marks the end of the Mesozoic: the famous Cretaceous-Tertiary boundary.[14]

---

13 See Note 2.

14 The term 'Tertiary' was coined in 1759 by an Italian geologist named Arduino, who used this name to describe relatively poorly consolidated and only slightly deformed rocks, while the underlying 'Secondary' rocks were simply more deformed and harder, and the 'Primary' basement exposed in some nearby mountains was even more severely affected. In 1833, Lyell subdivided the Tertiary, calling its earliest level the Eocene Epoch. After a number of different incarnations, the term Paleocene was introduced, which at first referred to the lower part of the Eocene and later became an epoch in its own right. As for the Cretaceous, the last period of the Secondary, it was introduced by Halloy in 1822 and takes its name from the chalk which often forms the strata of this age in northwestern Europe. In fact, we know today that the boundary between the Cretaceous and the Tertiary Periods, which as we will see is not easy to define nor often all that easy to observe precisely, is quite simply absent in the two regions where these periods were defined. Whether the corresponding strata were never laid down or were worn away later by erosion, this moment of geological time is not recorded there. The Cretaceous-Tertiary boundary is often known 'familiarly' as KT; the K refers to the first letter of Cretaceous in German ('Kreide'), so as not to confuse it with either Carboniferous or Cambrian (designated, respectively, as C and Є).

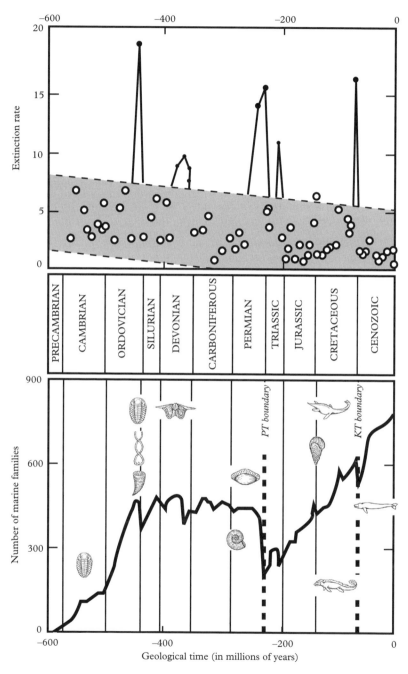

Figure 1.2
Changes in species diversity (actually illustrated by the total number of marine families rather than species) (bottom) and extinction rate (measured as number of families becoming extinct per million years) (top) as a function of time. (After David Raup and John Sepkoski.)

After this crisis, the diversity of species recovers very rapidly, and then grows more slowly for 30 Ma, recently achieving its highest levels since the beginning of Life on Earth. The great accidents are even more evident when we look at the extinction rate in relation to the number of families in existence at a given moment (Fig. 1.2, top). This rate of extinction undergoes rapid but relatively slight fluctuations around a mean value that regularly declines over time. Some of these fluctuations undoubtedly result from observational errors or uncertainties, but most of them merely reflect the "normal" rate of extinction (on the order of one family per million years) we discussed above. Against this "background noise" we see five peaks, which correspond to the five major crises mentioned earlier. According to Raup, long periods of profound boredom were thus interrupted episodically by brief moments of unfathomable panic. We may, moreover, wonder whether these moments differ from other more "normal" periods of extinction in some really fundamental way, or only in intensity. In the latter cases they would be part of a continuum, just like the "hundred-year flood" among all observed floods, or the "hundred-year earthquake" within the catalog of more "normal" quakes.

The unreliability of the sedimentary message

Paleontologists are anything but unanimous about either the duration or the nature of the great ecosystem upheavals. Geologists have been working on crisis scenarios for 150 years, and any successful version must be based on observations that are as clearly quantified, precise, and complete as possible. A mass extinction can be characterized by its duration, intensity (rate of extinction), and magnitude (diversity of affected groups). In estimating these parameters, we have to rely on the quality of the record of this entire history preserved in sedimentary rock. We will soon see that the foremost among the various contending scenarios offer very different pictures of these extinctions.

Paleontologists represent the lifespan of a species (or family, or group) by a line along the time axis. This stands for the thickness of the sediments in which this species was found within a stratigraphic section. Most of the time, these observations are incomplete,

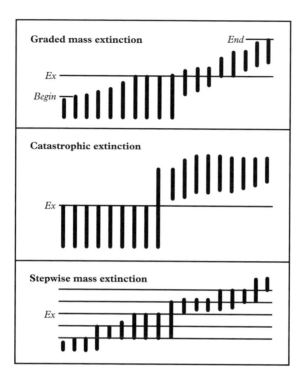

Figure 1.3
Various extinction scenarii (each vertical bar represents presence of a given species at a given level or time). (After P. Hut.)

and their interpretation is debatable. This explains why we have at least three possible scenarios for a great many boundaries between geological stages. First, there is a gradual scenario, in which species disappear and appear regularly one after the other (Fig. 1.3, top). This scenario is defended by the uniformitarians, who see it, for example, as the result of slow modifications (on the scale of tens of millions of years) in climate or in sea level. Then there is the scenario of an instantaneous, catastrophic extinction of numerous species, followed by a gradual reappearance of new life forms (Fig. 1.3, center). Finally, there is an episodic, "stepwise" scenario made up of a rapid succession of several events that are less intense than in the catastrophic scenario (Fig. 1.3, bottom).

Fossils and the strata containing them are, in fact, very haphazardly preserved. Discovering the last bones of a species at a certain level in the section of a formation by no means guarantees that this level really corresponds to an extinction. The larger the average size of the individuals in a species, the fewer of these individuals there will generally be: there are fewer elephants than mice, for example. In this regard the human race, as it proliferates across the globe, represents something of an exception. So fossils of large species are rare, and we can never be sure that more attentive study might not reveal them in more recent formations. A variety of sedimentary phenomena may complicate the picture still further. Erosion may pick up bones and redeposit them farther down along a channel, in younger layers. On ocean floors, burrowing organisms displace sediments across a certain thickness and thus may redistribute the microfossils they contain.

The accumulation of sediments is not a continuous phenomenon. The sedimentation rate may vary considerably and almost instantaneously, for example when a calmer sedimentation process is interrupted by a mud flow, landslide, or turbidity current. Sedimentation may quite simply come to a halt for a more or less extended period. Finally, erosion may completely erase an entire section of sediments from the record. So time is very irregularly recorded in rock.[15] A mere hiatus in sedimentation can make a phase of gradual extinction look like a mass extinction.

More subtly, incomplete sampling will make a sudden extinction look gradual. Very rare species may easily be "missed" by some observers and therefore, appear to have died out sooner than they really did. On the basis of the remarkable collection of ammonites gathered by Peter Ward at Zumaya in Spain, Raup showed how sedimentary hiatuses of various magnitudes may "mimic" both sudden and gradual extinctions. In brief, the discovery of a continuous sequence and a representative record is a long shot, and the sites of such discoveries become pilgrimage points for the international geoscientific community. For the KT boundary, repeated sampling

15 The incomplete and episodic aspect of sedimentation becomes very evident when one studies how the sedimentation rate varies as a function of the time interval over which it is measured. On the small scale, there may be numerous gaps, but while the sediment is being deposited the rates will be high. On the larger scale, the mean rate becomes lower and lower. The law linking these two quantities is of the 'power' type and characterizes the self-similar processes and fractal objects introduced by B. Mandelbrot: the distribution of gaps in the sequence appears the same on all scales. (Interested readers may want to refer to J. Gleick's book *Chaos: Making a New Science*, New York, Viking, 1987.) Typically, the mean sedimentation rate is in the order of one centimeter per thousand years, over an interval of one million years; but it rises to several meters per thousand years on the thousand-year scale.

has left the sections at Stevens (or Stevns) Klint in Denmark, Gubbio
in Italy, El Kef in Tunisia, and the Brazos River in Texas as full of
holes as a slice of Swiss cheese. It now seems that some of these
sections, considered continuous as recently as in the mid-1980s, are
often interrupted by cessations of sedimentation that had not been
detected at first. We must bear in mind these fundamental limitations of the quality of the stratigraphic record, which will have consequences not just for our interpretation of fossil distribution but
also for many physical and chemical indicators we will discuss below.
The upshot is a simple warning. It makes little sense to perform a
highly sophisticated and precise analysis of a sample's content of
iridium, carbon isotopes, or shocked minerals, or its magnetization
(measurements we will return to below) unless one has carefully
situated the sample in the formation and placed it within its sedimentary and stratigraphic context.

The last great crisis

The farther back we try to go in time, the more we feel the effects
of our geological myopia. Indicators become more and more frag-
mentary, and harder to decipher. So let us start with the least far-
off of these mysteries, the one where there has not yet been enough
time to eliminate all trace of the culprit – the KT crisis. What can
paleontologists tell us about this last great crisis that struck our
planet?

   To start with, what about the famous dinosaurs, those dragons res-
urrected from oblivion who continue to support astonishing waves of
advertising but also plainly still provide interest or amusement for a
great many people? Some experts say the last remains are clearly older
than the KT boundary, possibly by 200,000 years; others, that they
are 200,000 years more recent! But the "evidence," which comes from
Montana, is hotly contested. The fossils may have been displaced by
later geological events. The fossils of the largest of these animals are
rare, the picture of their disappearance very fuzzy in any case and,
in fact, differs from one continent to another. Sometimes a great
distance separates the last dinosaur bones from the first Cenozoic
mammal bones. It has not yet been possible to establish for certain
whether the last great saurians disappeared simultaneously. The

picture on which many paleontologists seem to be converging, however, is of a gradual decline in the diversity of dinosaur species over the last few million years of the Mesozoic, with an undoubted acceleration several hundred thousand years before the boundary. So far as the dinosaurs are concerned, we cannot (yet?) speak strictly of a sudden mass extinction.

Other terrestrial vertebrates were affected, among them the flying reptiles and the marsupials. But freshwater fish and amphibians, turtles and crocodiles, and snakes and lizards were almost untouched, and placental mammals, whose fate particularly concerns us since our ancestor was among them, survived. In the seas, one group of large reptiles, the mosasaurs, died out; over half the sharks and rays disappeared, but the rest lived on. In general, it was the larger and the more "specialized" animals that vanished, while the smaller ones and the "generalists" pulled through rather well.[16] Those with the broadest geographical distribution in the most varied environments survived better than the others.

The evolution of vegetation close to the KT boundary seems confused. Some experts speak of a gradual decline that started a few million years before; others emphasize the discovery, in North America, of an uncommon abundance of fern spores. These "opportunistic" plants are the first to recolonize a forest after a fire. They may mark the reconquest of a devastated world from which we know that many flowering plants, the angiosperms, had disappeared. Yet a few hundred kilometers farther north, in Canada, we find no further trace of this "fern peak," and the effects of mass extinction seem greatly reduced. The French paleontologist Eric Buffetaut stresses this selective, nonuniform aspect of extinctions in the continental context. To his way of thinking, a severe deterioration of climate or a simple size effect (the disappearance of the largest forms) cannot by themselves be a cause of extinction. The crocodiles, for example, which according to him are as sensitive to cold as the dinosaurs were, survived. Large crocodiles "made it" across the boundary, while many small marsupials did not. Noting that freshwater communities did not suffer too much, and that it was the large plant-eaters that disappeared, Buffetaut suggests that a crisis in the plant kingdom interrupted the food chain, thus wiping out the herbivorous dinosaurs and by consequence their carnivorous predators. Meanwhile the small carnivorous, insectivorous, or omnivorous vertebrates, and the freshwater organisms, whose food chains did not depend so heavily on the plant kingdom survived.

For his part, the American scientist Robert Bakker, an original and controversial specialist in dinosaurs, has long contended that a great number of these saurians, and particularly the largest and most active, were warm-blooded. So the comparison with crocodiles and other cold-blooded animals would no longer apply. Bakker believes that the extinction of his favorite animals was a prolonged event, quite simply caused by the low sea level at the end of the Cretaceous. This made it possible for the more mobile species, the more prodigal expenders of energy, to migrate over long distances, increasing the risk that they might succumb to diseases to which they were not resistant; by comparison, the smaller animals (among them our ancestors) and the cold-blooded species, being less mobile, would not have traveled far from their original habitat. This idea goes back to one of the fathers of the study of dinosaurs, Owen (1804–92), who was struck at the devastation caused by the introduction of bovine leprosy in Africa and at the adverse implications for kangaroos when rabbits were brought to Australia.

However, the continental paleontologic record by itself does not permit us to determine either the duration of the crisis, or its first causes. How can we evaluate the influence of changes in rock type and rock preservation (which may extend to a total absence of some periods)? How can we assess the local, regional, or global value of a given observation? How can we study scale in space and time or seek the cause of crises, whether fluctuations in climate or in sea level?

The marine environment, where sedimentation is generally more regular than in the continental context, offers more hope. The hard parts of the bodies of marine animals fall to the bottom and are rapidly covered. But 90% of geological sections from the KT boundary are incomplete and give the appearance of a single and abrupt mass extinction. Detailed analysis of the few very continuous sections (those with high sedimentation rates) where the rock and its fossils have been studied centimeter by centimeter yields a very different spectacle.

The marine invertebrates, such as the mollusks, do not furnish a very clear picture. Their diversity and abundance decline a few hundred thousand years before the boundary,[17] and then at the boundary itself. Some generalist species of simple morphology survive into the start of the Cenozoic. Ward's work with ammonites in the Basque country first showed a decline in species' diversity long before the boundary. But new fossils discovered in nearby sections a few years later have now attested the presence of some species of ammonites within a few meters of the boundary. And in the summer of 1996 the geochemist and astrophysicist Robert Rocchia even found a beautiful mold of an ammonite shell only 7 cm below the boundary! Today, Ward believes that a gradual extinction, caused by the slow drop in sea level at the end of the Cretaceous was followed by a final, abrupt extinction.

A catastrophe that lasted a hundred thousand years

In fact, the main part of our observations and interpretations of the KT crisis is now founded on the massive and apparently catastrophic extinction of almost all species of marine Foraminifera,[18] which make up plankton. In the late 1970s and early 1980s, paleontologists thought the most continuous sections would be found in the deepest sediments. Ocean cores, as well as sections outcropping on land, showed an abrupt succession of a carbonate mud rich in Cretaceous fossils, followed by a thin, dark layer of almost unpopulated clay in which the "first" little Cenozoic Foraminifera appeared. It is upon such sections, which some today think to be incomplete, that the Dutch paleontologist Jan Smit based his 1982 assertion that all species of planktonic Foraminifera (except one) had suddenly died out at the KT boundary. In fact it was realized that more complete

17 This boundary, considered synchronous on the global scale, is defined here by the geochemical observation of a peak in the concentration of iridium, a metal very rare in the Earth's crust; we will discuss it at greater length in the next chapter. But other defining factors are a drop in the carbonate content of sediment in favor of clay, and an anomaly in the carbon-13 isotope, which is usually associated with a massive oxidation of organic matter (whether living or fossil). We will discuss this below as well.
18 One-celled animals, 0.1 to 2 mm in diameter, floating in surface waters.

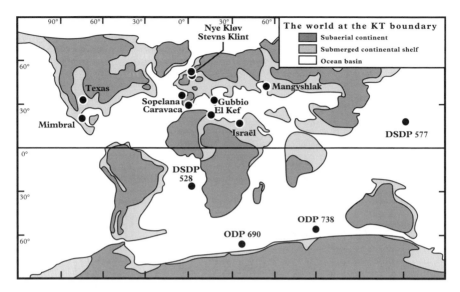

Figure 1.4
The world at the Cretaceous–Tertiary (KT) boundary, 65 Ma ago. Locations of the main sections, some quoted in the text, are indicated. (After Gerta Keller.)

series might be preserved in marine sediments deposited on the continental shelf, in comparatively shallow waters. There, a new stratigraphic and biological zone was discovered, which is quite simply absent from almost all the deep ocean sediment sections.[19]

Figure 1.4 shows the probable geography of the world at the end of the Mesozoic, with the emerged continents and submerged shelves, and the sites where the most complete (or, more accurately, the least incomplete) sections were discovered. Most of these names are now famous among the geoscientific community, and we will encounter them again and again. Gerta Keller's work at El Kef offers a fine example. There, this Princeton paleontologist devoted detailed study to the sequence of disappearances and appearances of nearly 60 different species of planktonic Foraminifera over a thickness of 5 m, representing several hundred thousand years on either side of the boundary (Fig. 1.5). Although nearly one-third of the species disappear at this level, an equal quantity disappear earlier, 25 cm

19 This is, for instance, the view of Gerta Keller
and her associates.

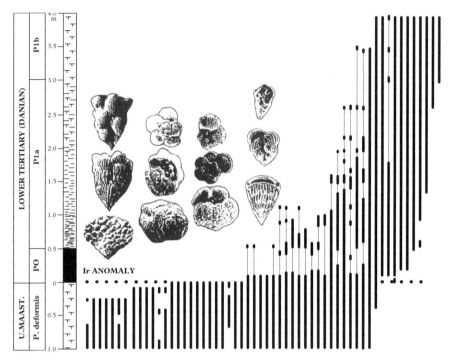

Figure 1.5
Species distribution in the El Kef (Tunisia) section (after Gerta Keller). Each vertical bar represents presence of a given species, some of which are illustrated. Geological stages, paleontologic zones, and sedimentary facies are given as columns to the left. The reference Cretaceous–Tertiary (KT) boundary, defined by the spike in iridium concentration, is indicated.

deeper down, and the rest in several stages above. The general look of these extinctions is reminiscent of the "stepwise" model we mentioned earlier. So the crisis that led to the disappearance of these species apparently was not instantaneous. But this is challenged by other paleontologists, for example Jan Smit.

Many tropical or subtropical species, with relatively large and delicately ornate skeletons, were the first to go, leaving room for smaller, simpler and hardier species (generalists). The characteristic increase in the total number of individuals of some species that survived the crisis, together with the systematic decrease in their size,[20] shows that this crisis began before the KT "boundary"[21] and continued

20 And their isotopic oxygen content, which makes it possible to distinguish species that lived at the end of the Cretaceous Period from those

from the start of the Tertiary Period.
21 See Note 17.

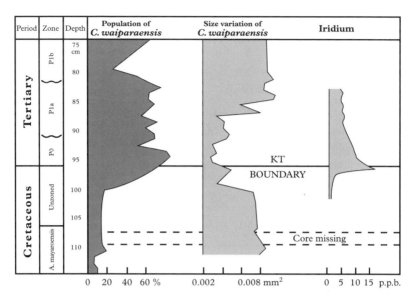

Period	Zone	Depth	Population of *C. waiparaensis*	Size variation of *C. waiparaensis*	Iridium

Figure 1.6

Stratigraphic distribution of population and size variation of microfossil *C. waiparaensis* across the Cretaceous–Tertiary (KT) boundary (after Gerta Keller). Iridium concentration is given on the right. The size decrease then restoration of this survivor species is clearly marked.

beyond it (Fig. 1.6). Thus it seems that the biological crisis began at least 100,000 years before this famous boundary and continued for about as long after. So longer-term events, spread out over millions of years and undoubtedly a function of climate, sea level or even, more simply, ongoing interactions among species, are overlaid by an abnormal period of less than half a million years, punctuated with phases that were more abrupt, but about which we cannot say for sure whether they lasted less than a day or more than a thousand years. At El Kef, one or two phases precede the boundary itself. At the Brazos River, the boundary is accompanied by no extinction at all but is preceded and followed by two rather sudden and intense events. The return to normal was particularly slow, and it seems that the ecosystems took more than 500,000 years to really recover. Marine extinctions are selective and affect deep-water and medium-depth species earlier and more completely than those that proliferate on the surface. At many sites where an extinction had seemed single and sudden, a large slice of time had in fact been "condensed" into a few millimeters or simply eroded away. Yet this global phenomenon, linked with a drop in sea level and a slower rate of sedimentation, is itself evidence of quite an exceptional event.

Where does all this leave us?

After a century and a half of patient and sometimes contradictory work, stratigraphers and paleontologists have provided convincing evidence of several extinctions of exceptional intensity. Can we still track down the culprit or culprits in these massacres? Where Lyell and Darwin saw only the conjoined effects of natural evolution, the immensity of geological time, and the capriciousness of the rock record, Buffon and Cuvier perceived catastrophes and named their suspects: changes in the environment, according to the one; the Flood, according to the other. Where does all this leave us?

Let's go back to the late 1970s. For many researchers, the KT crisis is certainly a remarkable event, but no one yet seems able to put a value on its duration with a precision greater than a million years, still less determine its causes. So we must start by studying its details. Very well, we're already on our way: part of the answer, which will mark the beginning of an exciting period, is being assembled somewhere in Italy, a few kilometers north of a charming hill town in Umbria.

2

An asteroid impact

Between the Jurassic and the Cenozoic,[1] several thousand meters of limestone were laid down in a shallow sea where the region of Umbria in Italy is today (see Fig. 1.4, p. 19). Compacted, transformed into hard rock, then folded in the uplift of the Apennines, today this stone forms the hills from which the ochre or pink blocks of *scaglia rossa* are quarried to build the beautiful buildings of the town of Gubbio. The geological sections that line the roadways leading away from Gubbio have long been known to geologists. It is here that the first Foraminifera were identified, in the 1930s. Some of these sections have been almost literally honeycombed by paleontologists and paleomagneticians, foremost among them the Scotsman William Lowrie and a young American geologist named Walter Alvarez.

Father, son, iridium, and impact

One evening in 1977, Walter Alvarez brought a small specimen the size of a packet of cigarettes, from the Gubbio section, to his father Luis, the famous Berkeley physicist and Nobel laureate. The geologist son pointed out to his physicist father the sequence in which several centimeters of white limestone were followed by a thin layer of darker clay 2 cm thick, and finally by several centimeters of reddish limestone. Under the magnifying glass, they could see Cretaceous Foraminifera in the white strata, but nothing in the clay. Above this began the Cenozoic layer, and with it the slow resumption of life. Luis Alvarez was holding in his hand a small piece of evidence of the end of the Mesozoic, possibly contemporary with the last dinosaur (Fig. 2.1). In his autobiography,[2] he would write how this moment

1 From 180 to 30 million years ago.
2 *Adventures of a physicist*, New York, Basic Books, 1987. Since the French version of my book was published (1995), Walter Alvarez has produced his own, first-hand and long awaited account: T. rex *and the Crater of Doom*, Princeton, Princeton University Press, 1997.

Figure 2.1
The Cretaceous–Tertiary (KT) geological boundary at Gubbio (Italy). A thin dark clay bed, a few centimeters thick, separates the Cretaceous limestone beds (lower right) from the Tertiary ones (upper left). Microscope observation of fossils reveals the mass extinction and evolutionary turnover (Robert Rocchia).

was the birth of his interest in paleontology, a discipline which he had evidently disdained somewhat until then, and whose practitioners he would continue to hold in scant regard even afterwards.

The two Alvarezes wondered what length of time the mysterious layer of clay recorded. Luis then had the idea of measuring this duration, perhaps very brief and dating from so very long ago, with a highly original chronometer: the deposition of exotic material from the incessant rain of micrometeorites falling to Earth. Micrometeorites are in fact rich in certain chemical elements, particularly those of the platinum family, that are very rare in the Earth's crust. Among these elements iridium (Ir), number 77 in the periodic table, was the easiest to measure by the new technique of neutron activation. Two of Luis's colleagues at the Lawrence Livermore Laboratory, Frank Asaro and Helen Michel, were experts in this technique, which consists of bombarding a specimen with a neutron flux that turns the iridium radioactive. The level of this induced radioactivity can then be measured. Luis Alvarez thought that by measuring the iridium profile in the specimen his son had brought him, and on the assumption that this iridium derived from the steady rain of micrometeorites, he would be able to measure the time that had elapsed during the deposition of the layer of clay. And thus he unwittingly reinvented a technique for measuring sedimentation rates that had first been proposed back in 1968.

The obtained concentrations were minuscule, the products of an analytical tour de force: far from the Cretaceous-Tertiary boundary, they have been established at a few tenths of a part per billion (p.p.b.). But in the clay layer, they attained 9 p.p.b., a value 30 times greater. Abnormal values were also found as much as 15 cm above the clay layer. Now, in the Earth's crust, the natural concentration of iridium is a thousand times less, rarely exceeding a few hundredths of a part per billion! Very excited at their discovery, which implied quantities of iridium far greater than those that would have been deposited by a simple rain of micrometeorites even over several million years, the team immediately began looking for an abnormal event of extraterrestrial origin. The first "culprit" they thought of was the explosion of a supernova in the vicinity of the solar system. But the absence of plutonium-244 quickly ruled out that hypothesis. In the year following the discovery, numerous scenarios were proposed, tested, and rejected. Finally one of the Alvarezes' colleagues, a Berkeley astronomer, suggested an asteroid impact. Some meteorites do contain iridium concentrations in the order of 500 p.p.b., 50,000 times greater than in the Earth's crust. Assuming that the abnormal layer of iridium would be present all over the Earth's surface, and knowing its thickness and concentration, it would be possible to calculate the total mass of iridium so suddenly introduced 65 million years ago. Working from the content of this metal in various types of meteorite, they could estimate the approximate size of the extraterrestrial object: 10 km in diameter. At the phenomenal speed of the impact, this would imply a release of kinetic energy equivalent to 100 million megatons of TNT, ten thousand times greater than the planet's entire nuclear arsenal![3] The impact hypothesis had been born.

---

3 The unit often used to measure the energy of an impact is a million tons (or megaton) of TNT. In the International System of measurements, this is equivalent to 4×10^{15} joules. So an asteroid 1 km in diameter corresponds to an energy of 100,000 megatons, and the energy of the Alvarezes' asteroid is equivalent to 100 million megatons.

The painstaking work of the Italian and American teams at Gubbio in the late 1970s, which resulted in this scientific bombshell, is worth recounting. For it is also an important step in the construction of a global scale for geological time.

Magnus magnes ipse est globus terrestris[4]

The Earth behaves like a magnet. Indeed, it has its own magnetic field. It is this magnetic field that provides the orientation for compasses, which themselves are small magnets, and are still sometimes used today for direction finding when the sun is not visible. The blue end of the compass needle (its north pole) is attracted by the south pole of our planetary magnet, which is situated not far from the geographic North Pole[5] (about 1000 km from it, somewhere in Canada). But this was not always so. Around 800,000 years ago, the magnetic *north* was located near the geographic North Pole. Had there been compasses in those days, and had they been colored the same way as ours are, the blue ends of their needles would have pointed south. Indeed, over the past 100 million years the Earth's magnetic field has reversed hundreds of times, very irregularly. The discovery of these reversals goes back to the work of the French scientists Brunhes and David early in the twentieth century but was only accepted by the scientific community in the 1950s.[6]

When a lava cools, the tiny crystals of certain iron oxides such as magnetite are magnetized with the same direction and orientation as the ambient magnetic field, and this magnetization may be preserved almost indefinitely as long as the rock is not altered or reheated. A specimen of basalt is thus a regular little permanent compass. In the 1960s, new generations of instruments had become available that were capable of very sensitive datings and magnetic measurements: geochronologists had new mass spectrometers, able to measure very small concentrations of the elements potassium and argon, while the

---

4 "The earth itself is a great magnet," said William Gilbert, physician to Queen Elizabeth I of England. See Chapter 3.

5 By common error, people say that the (north, blue) needle of the compass points to the north magnetic pole.

6 See J.P. Valet and V. Courtillot, Les inversions du champ magnétique terrestre. *La Recherche*, 23, 1002–1013, 1992. A popular account in English is W. Glen, *The road to Jaramillo: Critical Years of the Revolution in Earth Science*, Stanford, CA: Stanford University Press.

paleomagneticians had a high-speed spinner magnetometer. During the nineteen-sixties, two teams – one American, one Australian – were able to work out the first time scales for the reversals of the Earth's magnetic field, by carefully dating many lava specimens from all over the globe and measuring their magnetization.

From oceanic magnetic profiles . . .

At the same time, geophysicists recorded the fluctuations of the magnetic field with another kind of magnetometer (called a "fluxgate magnetometer") towed behind the boats that more and more commonly traversed the seas in the attempt to unravel the mysteries that lay hidden beneath several kilometers of salt water. The magnetic profiles thus obtained have astonishing characteristics. The magnetic anomalies alternate in parallel, symmetric positive and negative bands on either side of the great submarine chains of the mid-ocean ridges, and can be traced for thousands of kilometers. In 1963, in two independent articles that have by now become famous, the Canadians Morley and Larochelle and the Britons Vine and Matthews had the brilliant idea of interpreting these profiles in the light of the nascent theory of sea floor spreading and the first magnetic reversal scales (Fig. 2.2).

The molten material extruded from the Earth's mantle cools along the axes of the mid-ocean ridges to form the basaltic oceanic crust. The direction of the magnetic field is fixed in the rock at that moment. Like a double conveyor belt, the solidified crust then moves away and new crust takes its place along the axis of the ridge, where in turn it sets as it cools. When the Earth's magnetic field reverses, it imprints a symmetric double sequence of bands on the ocean floor, alternately magnetized first in one direction and then in the other. Then all one has to do is recognize the characteristic appearance of these bands (like a kind of bar code) and measure their distance from the axis of the ridge to determine the rate of ocean expansion and the accompanying passive drift of the continents. Though sometimes less than 1 cm per year, this rate may be as great as 20 cm annually. To give an idea of its order of magnitude, the North Atlantic Ocean grows wider by approximately a man's height during his lifetime.

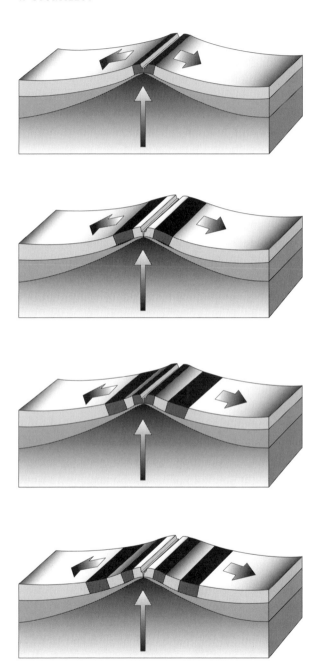

Figure 2.2
Formation of magnetic anomalies, alternatively positive and negative, on the oceanic crust on both sides of a mid-oceanic rift. The oceanic crust records field reversals as a magnetic tape, in a symmetrical pattern, when rising volcanic material cools and drifts away.

Direct measurements of the ages and magnetic polarities of the lava from ocean islands or on continents had furnished the sequence of reversals over only about 4 million years. By making a daring extrapolation from a width of 100 km on either side of the ridge (5 million years at a rate of 2 cm per year) to the width of the entire ocean, it was possible to reconstruct the history of magnetic reversals back to the age of the oldest ocean floors,[7] about 160 million years (the mid-Jurassic Period).[8] Since the 1970s, sediment samples and cores drilled from the oceanic crust have made it possible to pin down the sequence of these reversals, both by geochronologic (using potassium–argon) and biostratigraphic methods (using fossils). From now on the time reference in geology will be a scale subject to three constraints: fossils, magnetic reversals, and absolute age (Fig. 2.3). In addition, changes in sediment composition related to the Milankovic climate cycles (see Chapter 1, Note 11) leave a cyclic imprint that can be used to provide even finer timing. This forms the branch of the geosciences known as cyclostratigraphy.

. . . to magnetostratigraphy

The sediments, too, record the magnetic message, albeit in a very different way. Magnetized particles, grains resulting from the breakdown and transport of continental rock or from biological activity,[9] are deposited in sediment and orient themselves in the direction of the ambient magnetic field. After the sediment has expelled the water it initially contained and has become rock at the end of the physical and chemical processes known collectively as diagenesis, this very weak magnetization (far weaker than in the lava from volcanoes or in the oceanic crust) may be preserved.

What Lowrie, Alvarez, and their colleagues were looking for in the *scaglia rossa* was precisely the sequence of reversals that had

7 These floors have been dated absolutely from lavas sampled as part of oceanic projects, and relatively using the sediments covering them.
8 It has also been possible to reconstruct the history of the opening of these oceans themselves. The fact that the oldest oceans are 25 times younger than the Earth (4.5 billion years) was in itself an important discovery.

9 Some one-celled animals synthesize tiny crystals of magnetite; these allow them to orient themselves in water and find the bottom, which is their food source. See J.-P. Poirier, *Le minéral et le vivant*, Paris, Fayard, 1995.

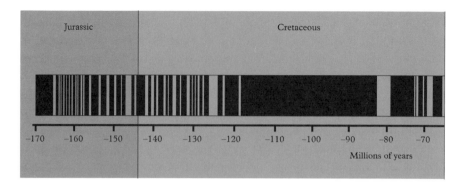

Figure 2.3
The magnetic reversal time scale, since the Jurassic Period, 170 Ma ago. Normal periods are in black, reversed ones in white. The names of some magnetic "chrons" (chronozones) are given (we are still in the Brunhes).

already been discovered "horizontally" across hundreds of kilometers of profiles from sea floors, but this time occurring "vertically" in sedimentary rock and only across a few hundred meters of a section.[10] The development of new, even more sensitive magnetometers[11] would make this possible. In 1977, the Italian and American teams published a remarkable series of five articles in the bulletin of the Geological Society of America. Almost without gaps, the authors clearly found the readily recognizable sequence of magnetic polarity intervals that had made it possible to found the field of plate tectonics 15 years before. The long interval of magnetic polarity analogous to our current polarity (and for that reason called "normal") that prevailed in the Cretaceous Period, some 85 to 120 Ma ago, was succeeded by a more and more rapid alternation of "reversed" and normal periods, which could be paired with the duly catalogued and numbered "chronozones"[12] of the marine profiles. This made it

10 This difference reflects the difference between the rate of sea floor spreading (a few centimeters a year) and the mean rate of sedimentation (a few centimeters per thousand years).
11 These are cryogenic magnetometers, using superconducting pickups submerged in liquid helium at 4 degrees Kelvin (-269 °C). They make it possible to detect magnetic fields of one millionth of a millionth of a tesla (10^{-12} T), 50 million times weaker than the magnetic field of the Earth at its surface (about 5×10^{-5} T). This latter field in its turn is a hundred thousand times weaker than the field produced by some magnets used in particle physics (about 5 T). To give you a more tangible idea of these relations, the field of a small "pocket" magnet is on the order of 10 microteslas (10^{-5} T) at a distance of 20 cm. It is only 80 nanoteslas (8×10^{-8} T) one meter away (the decrease varies as the cube of the inverse of the distance).
12 Chronozones are often referred to as chrons.

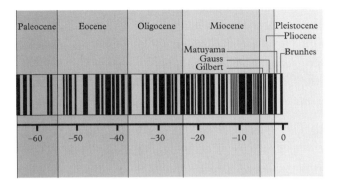

possible, in particular, to show that the boundary between the Cretaceous and the Cenozoic[13] is located within a reversed period known by the unglamorous name of "29R."[14]

The limestone beds corresponding to this period extended across about 5 m of a stratigraphic section and represented about half a million years. Based on these observations and the thickness of the thin stratum of clay at the KT boundary at Gubbio, Dennis Kent, of the Lamont Doherty Geological Observatory near New York where Lowrie and Alvarez were also members at the time, was the first to suggest that the events responsible for mass extinction had lasted less than 10,000 years! This hypothesis, almost incredible at the time, would play an important role in orienting Walter Alvarez's research and convince him that a major discovery was within reach of his geologist's hammer.

An ecological disaster

Published in the journal *Science* in 1980, the hypothesis of an asteroid impact on Earth withstood its first trials. The proposed scenario for the mass extinction of species was the "impact winter." The asteroid, pulverized and vaporized, would have shot into the atmosphere

13 The KT boundary, see Chapter 1, Note 14.
14 Very roughly, this is the 29th major interval
of reversed polarity (neglecting very short events). See Figure 2.3.

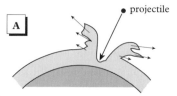

Direct ejection during atmospheric passage

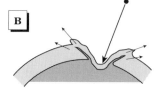

Entrainment by fast solid ejecta

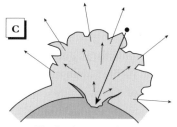

Ejection by vapor plume

Figure 2.4
Ejection of atmospheric
gases (light gray) and solid
crust (darker gray) during
early impact of a celestical
projectile (after R. Vickery
and J. Melosh).

a mass of terrestrial matter (Fig. 2.4) several tens of times greater than its own mass (amounting to some 10^{14} kg). This dust would take months or even years to settle again. It would have shut out the sunlight, arresting photosynthesis and causing a very long, hard winter. The disappearance of plants would break the food chains, and the carnage would begin. (In 1981, a very similar scenario was constructed to describe what would happen in a generalized nuclear conflict; with heavier media coverage than the impact winter, the "nuclear winter" would enter into the collective awareness.)

More precisely, on impact, the asteroid would essentially have dug a hole in the atmosphere, and then in the earth's crust. The released energy and the ejected products, particularly the tiny fragments that would have risen in a ballistic trajectory and returned to the atmosphere, generating considerable heat, together would have produced abundant quantities of nitrogen oxide.[15] Fires, which we will discuss

15 By high-temperature reaction between the
air's nitrogen and oxygen, heated by the fireball
from the shock and by the settling dust.

below, would likewise have produced not only vast quantities of soot but also more nitrogen oxide. Combining with water, the latter would produce nitric acid aerosols, capable of destroying the protective ozone layer, and acid rains that would damage vegetation and even dissolve the calcareous skeletons of microorganisms living in the surface layers of the ocean.

In 1987, shortly before his death, Luis Alvarez was clearly very pleased with 'his' theory (as who would not be?): he insisted in his autobiography on the evidence of the theory's correctness and the numerous predictions it had made possible (and that he believed verified at the time). Some of his emulators would not hesitate to try to make his opponents the laughing stock of scientific meetings. Luis Alvarez once described paleontologists who were too slow to accept his theory as "stamp collectors." Contrasting facets of a very colorful personality.

It's raining meteorites

The article in *Science* would provoke impassioned reactions, whether acceptance or rejection, and relaunch research on the Cretaceous-Tertiary boundary on an unprecedented scale. An interdisciplinary effort, it would generate sheaves of publications and lead to meetings where paleontologists would rub shoulders with astrophysicists, geochemists, geophysicists, statisticians and many others. Since 1980, more than 2000 papers have been published on the subject. This book (and many others) would never have come into being without that founding article.

Among the opponents, some decried an appeal to a "deus ex machina," and rejected any extraterrestrial process. At least in appearance, this was a resumption of the war between Lyell and Cuvier, the "uniformitarians" versus the "catastrophists." But to a large degree the debate was off the mark. There is no longer any doubt that impacts of extraterrestrial objects have been of great importance in the history of the Earth. The Moon, an inert body without plate tectonics, still bears the marks of the great bombardments that shaped it during the first billion years of existence of the solar system: the great craters, like Copernicus and Tycho, remain as the evidence (Fig. 2.5). Our natural satellite itself is undoubtedly

Figure 2.5
The Moon shows scars from intense early bombardment by asteroids and resulting lava
overflow.

the product of a gigantic impact at the time when the Earth was forming. The Earth has undergone the same bombardment, but erosion and the return of plates to the mantle have erased its traces.[16] Some experts believe the emergence of Life on Earth became possible only after the main phases of this bombardment were completed, a little more than 3500 Ma ago. At any rate, the impacts became rarer and rarer, and smaller and smaller; as the planets formed, in effect they "vacuumed up" the largest objects. But however attenuated, the cosmic bombardment has still not stopped. Every 30 microseconds, the space shuttles suffer the impact of a micrometeorite one micrometer in diameter (1 μm; micro means one millionth). A shooting star, corresponding to the fiery glow of a grain 1 mm in diameter, crosses the sky every 30 seconds. On average, one meteorite 1 m in diameter falls to Earth every year, and it is estimated that the impact of an object 100 m in diameter, like the one

16 This is the phenomenon of "subduction," which, for example, is responsible for the earth- quakes and volcanoes of the "Ring of Fire" around the Pacific.

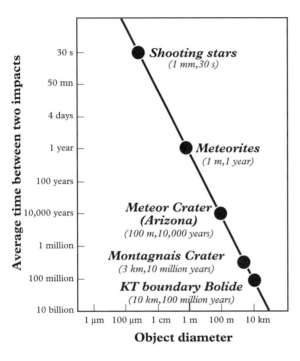

Figure 2.6
Average time between asteroid impacts as a function of object diameter. Note that the bottom line is older than the age of the Earth (4.5 billion years).

that blasted the famous Meteor Crater in Arizona, occurs an average of once every 10,000 years. By extrapolation, this scaling law (Fig. 2.6) would seem to indicate that a meteorite the size of the one the Alvarezes proposed falls to earth once every 100 million years.

Many asteroids, originating from the belt located between Mars and Jupiter, as well as some heads of comets deflected from their reservoir beyond the orbit of Pluto, follow orbits that may one day encounter the Earth. About a hundred asteroids with diameters of more than 1 km have been observed. Such discoveries have become more frequent under the influence of the *Science* article of 1980. There are an estimated 1000 of these objects in all. Astronomers have calculated the mean probability of impact at four in a billion per year, meaning that a meteorite 1 km in diameter, capable of creating a 10 km crater, falls to Earth every 250,000 years, and that an asteroid fragment 10 km in diameter, creating a 100 km crater, is likely to fall once every 300 million years. The impacts of the heads of comets, less dense but faster, might be more frequent and more devastating. The fall of the 21 pieces of comet Shoemaker–Levy onto

Jupiter in July 1994 offered an extraordinary example, fortunately at a very safe distance. The giant planet long bore marks of that impact. A comet head 10 km in diameter arriving at a speed of 30 km per second would create a crater 150 km in diameter on Earth. And this is thought to happen an average of once every 100 million years.

We owe some of these observations, as well as the scaling law derived from them, to Gene Shoemaker, the great impact specialist and the discoverer, with his wife and a colleague, of comet Shoemaker–Levy in 1993. In fact, it seems that many meteorites of substantial size explode or burn up at high altitudes, out of sight. Certainly the total number of meteorites that encounter the upper atmosphere of the Earth is greatly underestimated. In January 1994, the US Secretary of Defense released a list (hitherto secret) of 136 explosions that occurred between 1975 and 1992, detected by early warning satellites cruising at an altitude of 36,000 km. Each year there are nearly ten explosions with an intensity equivalent to 500–15,000 tons of TNT. Each of these meteorites must be around 10 m in diameter, weighs more than 1000 metric tons, and arrives at speeds in the order of 50,000 km per hour. So it seems that we must revise upward the number of objects whose trajectories intersect that of the Earth, but not necessarily the number of objects that reach the surface and are capable of digging a crater there.

Estimates of impact frequency are, in any case, very uncertain. They may be off by a factor of two (or even more), and it is risky to extrapolate them over durations that exceed by many powers of ten the period to which our observations apply. But it should be recalled that the impact of one asteroid 10 km in diameter every 50 to 500 million years is entirely plausible. The problem is to determine the climatic and ecological consequences of such an impact, and if possible to identify its traces with certainty. Luis Alvarez concluded from these probability calculations that an impact had *certainly* occurred during the past 100 million years, and that he had found the proof of it in the Gubbio iridium.

We know today that the underwater Montagnais Crater, off Nova Scotia, some 50 km wide and 50 million years old, had no effect on the diversity of species, even on a merely regional scale. And the twin craters of Kara and Ust Kara at the edge of the Arctic Ocean, one 65 km and the other more than 80 km in diameter, had hardly

any consequences when their meteorites fell to Earth some 75 million years ago. This does not mean that there were no catastrophes
at the level of individuals and populations; but at the scale of species
and families, the rate of extinction did not significantly exceed its
normal level (the background noise of extinctions). We only need to
recall that 2000 km^2 of forest were turned to charcoal when the
Tunguska comet exploded in Siberia in 1908, at an altitude of 10
km, to imagine what the surroundings would be like after a really
large impact!

Iridium and osmium

The 1980s would see an accumulation of evidence in favor of the
impact. First of all, the abnormal level of iridium was found at more
than a hundred different sites distributed all over the Earth's surface, sometimes in concentrations even higher than those at Gubbio.
The geological sections involved correspond to very different deposition environments, both oceanic and continental. Other geological
boundaries corresponding to other extinctions have also been
studied. After more than ten years of research, it seems that very
few other abnormal levels have been detected. None of the major
boundaries earlier than the KT includes a concentration of more
than a few tenths of a part per billion, and in all cases purely terrestrial sources or concentration processes could be responsible:
an accumulation of sulfides, phosphate nodules, rocks from the
upper mantle, a concentration of iridium in sea water due to bac-
teria, and so on. A layer richer in iridium, but with no connection
to extinctions, was found in the Jurassic rocks of the Alps by Robert
Rocchia and his associates from Gif-sur-Yvette, while the Alvarez
team has reported a layer 34 million years old in the Caribbean, corresponding to the extinction of five species of radiolarians, close to
the Eocene–Oligocene boundary. But so far, the iridium anomaly at
the Cretaceous-Tertiary boundary remains virtually (and astonishingly) unique.

Other elements from the same family as iridium, such as ruthenium and gold, also seem abnormally concentrated at the KT
boundary. In geochemistry, absolute concentrations of elements are
extremely variable. The concentration ratios tend more to charac-

terize the various "reservoirs" (or "sources") involved. An even greater degree of discrimination results when one uses isotopes[17] of a single element or of elements linked together in a chain of radioactive decay, for example, the isotopes osmium-186 and osmium-187. The rocks of the Earth's crust are richer than meteorites in rhenium; its radioactive isotope rhenium-187 decays into osmium-187. So the isotopic ratio of osmium (187/186) is higher in rocks of Earth origin than in those from meteorites. Jean-Marc Luck and Karl Turekian, both of them at Lamont at the time (1983), found very low values for this ratio in the layers at the KT boundary. Although clearly leaning toward the impact theory, these authors themselves pointed out that such low ratios were also compatible with the composition of the rocks of the Earth's mantle located below the crust. Could there be another scenario?

Spherules and shocked quartz

In 1981, Jan Smit discovered concentrations of tiny spheres in a number of sections from the KT boundary. These spheres were variable in composition but often basaltic and all more or less altered. A meteorite has such energy that it melts the rocks of the crust where the impact occurs and disperses them in ballistic trajectories, sometimes over very great distances. The little droplets, known as tektites, take characteristic shapes as they cool in the atmosphere. We find layers rich in splendid tektites in southeast Asia, and in marine sediments off Australia, dating from a few tens to a few hundreds of thousands of years ago. Although the impacts have not always been identified, no one doubts that extraterrestrial bodies are at the origin of their formation. Other tektites, called moldavites, have been found around the Ries impact site in Germany, and still others off the Ivory Coast. These seem to come from a nearby crater. So Smit proposed that the spherules of the KT boundary should be viewed as the altered remains of microtektites, supplementary evidence of an impact, and he has suggested that this impact must have occurred on an oceanic crust, the source of these droplets of molten basalt.

17 See Claude Allègre, *From stone to star*, Cambridge, MA, Harvard University Press, 1992.

In the early 1980s, Bruce Bohor, of the US Geological Survey was the next to look at numerous sites where the KT boundary is exposed. He discovered tiny mineral grains, particularly of quartz, smaller than 1 mm in diameter that seem to have suffered a shock of extraordinary violence (Fig. 2.7a). When we look at a thin slice ("section") of these grains under the microscope, at a magnification on the order of 1000, we see a great many little stripes that correspond to microscopic defects. Several families of these stripes, each parallel to a well-defined crystallographic direction, permeate these grains through and through. Now, such defects were already well known to petrographers, who had observed them in specimens from the sites of nuclear explosions or undisputed meteorite impacts, and in specimens shocked by the impact of a projectile in the laboratory.

When a grain is traversed by a brief shock wave at a pressure in excess of several billion pascals (i.e., several tens of thousands of atmospheres), thin layers of quartz (1 μm thick, or even less) lose their crystalline structure and are transformed into amorphous glass, whose refractive index[18] is a little lower than that of "normal" crystalline quartz. These microscopic optical contrasts – along with a special type of glass called "diaplectic," a very high-pressure form of quartz called stishovite, and "shatter cones," macroscopic striped conical fractures in a fan or horsetail shape – are very widely held to be indisputable indicators of an impact.[19]

However, other shocked quartzes were soon discovered in Austria, Bohemia, and Scandinavia at geological sites and for ages where the presence of an impact is far from accepted. Neville Carter believes he has observed structures of the same type in minerals associated with a gigantic prehistoric volcanic eruption at Toba, on the island of Sumatra. So observation under the light microscope alone is not enough to provide a reliable description and interpretation of these "planar defects" observed in grains. Since 1990, Jean-Claude Doukhan and his students at the University of Lille have been conducting a far more detailed analysis under transmission electron microscopy, which allows magnifications in excess of 100,000 and yields images of exceptional clarity and detail (Fig. 2.7b). Hidden

18 A measure of the ability of the material to bend a ray of light.

19 The set of transformations undergone by the rock is called "shock metamorphism."

(a)

(b)

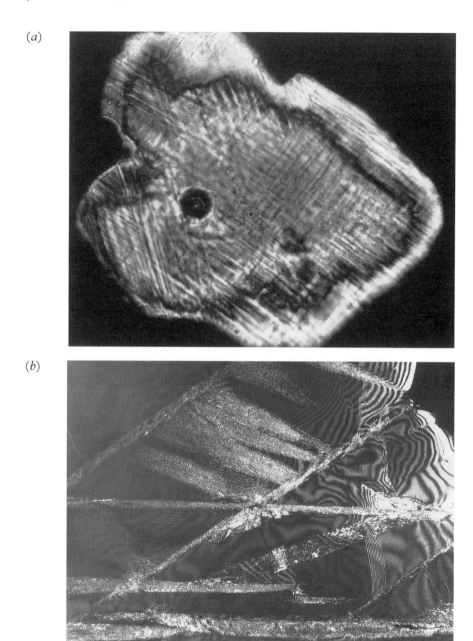

Figure 2.7
Shocked quartz grains from the KT boundary sediments. (a) Seen under the optical microscope in polarized light (figure is 1 mm wide) and (b) under the transmission electron microscope with much higher resolution (figure is 5 μm wide). Shocked quartz carries characteristic planar defects, seen here as dark lines, that correspond to glass lamellae and crystal twins resulting from brief passage of a strong shock wave (J. C. Doukhan).

beneath the apparent simplicity of the stripes observed under the optical microscope, they found a whole family of very different microstructures: "Brazil twins," glass lamellae, alignments between dislocations or bubbles of fluids, and more. Doukhan compared specimens from the KT boundary with quartz from the recent impact that formed the Ries crater in Germany, and with quartz shocked experimentally at more than 15 billion pascals. He indeed found the same twins and the same parallel glass lamellae, confirming that only a dynamic compression such as the kind that can occur in an impact seems capable of generating these characteristic defects.

Zircons and spinels

In association with Tom Krogh, of Toronto, in 1993 Bohor discovered that minuscule grains of zircon in a section from Raton Basin, in the USA, also presented signs of shock. Zircon, or zirconium silicate, is a semiprecious stone highly resistant to heating and alteration that incorporates small quantities of uranium at the time of its crystallization. The uranium–lead system is one of the most effective dating tools known to geology.[20] It presents the peculiarity of having two pairs of parent and daughter isotopes, uranium-236 and uranium-238, which decay in long chains to arrive at the stable isotopes lead-207 and lead-206. This provides two independent measurements of age (or in fact three, if we add thorium-232 and lead-208). This peculiarity, in association with extraordinarily precise analytical methods, makes it possible to determine whether specimens weighing a millionth of a gram, in which a millionth of a millionth of a gram of lead has been detected, underwent a perturbation after their first crystallization that would have partly reset their chronometer to zero![21] Therefore, these zircons are among the most ancient and most durable mementos of terrestrial events. The geochemist Claude Allègre gave one of his articles (unpublished) a title like a detective thriller: "Old zircons never die." The grains that

20 See Note 17.
21 A very substantial thermal perturbation may in fact erase the isotopic ratios of the elements trapped in the minerals of a rock and reset them to the surrounding equilibrium, leaving the impression that they have just been formed.

Krogh and Bohor observed come from rocks over 400 million years old, which some 57 million years ago (with an uncertainty of 4 million years) underwent not merely a shock but also a considerable thermal event. Krogh interprets this latter date as that of the KT boundary impact[22] and suggests that the traces should be sought in a region where at that time there was an outcropping of a small[23] zone of the continental basement containing far older rocks.

The indications multiplied. Jan Smit and Glen Izett, of the US Geological Survey, discovered small, well-crystallized minerals generally measuring a few micrometers across from the family of spinels formed from oxides of both iron and magnesium and known as magnesioferrites. These spinels are astonishingly rich in nickel (as much as 5%) and are already known from rocks linked with a meteorite impact. Robert Rocchia, Eric Robin, and their colleagues from Gif found high concentrations of these nickel-bearing spinels (far more abundant than grains of shocked quartz) in the El Kef section across a thickness of only a few millimeters, far thinner than the one over which the distribution of iridium extends. They see this as the trace of a single, abrupt extraterrestrial event that must have lasted less than a few centuries. They have no doubt that the nickel content and degree of oxidation of the spinels at the KT boundary demonstrate an extraterrestrial origin. But while these authors see magnesioferrites as the product from the condensation of the vaporized residues of a meteor in the atmosphere, Stan Cisowski, of the University of Santa Barbara, was impressed by their similarity to the particles dispersed by the wind during the natural or artificial combustion of fossil hydrocarbons, such as oil shales or petroleum. Thick layers of oil shales may have been exposed in abundance by the withdrawal of the seas at the end of the Cretaceous Period and could have caught fire spontaneously: the traces of such fires from other ages have been found in Israel and California.

---

22 However, it must be noted that the age of 57 ± 4 million years does not match that of 65 million years for the KT boundary. It is at least 4 million years short – no small amount. Krogh attributes this difference to a mixing of grain populations of different ages, or to a loss of lead.

23 Small, because the ages of zircons generally vary considerably over distances of 1 km or even 1 m, yet here they are very homogeneous. They come from sites in Haiti, Mexico, and Texas. The crater site can't be far away. We will return to this point in Chapter 8.

A general conflagration?

The existence of great fires at the KT boundary also seems to be attested by the presence of soot and natural wood charcoal discovered at Raton Basin in 1984, and later in Denmark, Spain, and New Zealand. High concentrations, of several milligrams per square centimeter, are associated with iridium but also with abnormal levels of arsenic, antimony, and zinc, which are of terrestrial origin. The signature of the carbon-13 isotope that forms this soot is similar to that of the charcoal from natural organic molecules synthesized by plants. Ed Anders and Wendy Wolbach, of Chicago, thus estimate that almost all living matter, the biomass, burned. According to these partisans of the impact theory, part of the Earth's vegetation was ignited by the fireball generated by the shock and by the thermal radiation of the particles that subsequently fell back down through the atmosphere and thus underwent considerable heating. The soot from these fires, added to the dust mobilized by the impact, would have prolonged the darkness and aggravated the resulting cold. Some of the remaining vegetation would have died, and its remains would presumably also have caught fire, this time under the action of lightning. On a similar scale, combustion releases carbon monoxide and organic toxins (such as dioxin) in such quantities that these products may induce mutations. The production of carbon dioxide would likewise have increased and contributed over the longer term to severe warming owing to the greenhouse effect, which would come on the heels of the impact winter. One almost wonders how any species at all, especially our presumed common ancestor *Purgatorius*,[24] could have survived this hell.

For almost 15 years, in successive stages, the arguments in favor of an impact seem to have accumulated inexorably. Like a good lawyer (or a bad one?), I have tried to present them as persuasively as possible. Moreover, the majority of the researchers in the disciplines concerned, particularly in the USA, seem to be convinced today by this hypothesis. But in fact, braving increasingly vehement criticism as the theory of the extraterrestrial object consolidated into

---

24 This name, which seems too good to be true, belongs to a small, rat-sized, insect- and fruit-eating mammal – the first, in fact, to have eaten fruit – whose remains were found at the Purgatory Hills site in the USA.

a new paradigm, other researchers found difficulties and contradictions and proposed another scenario. Contrary to Bohor's claim, the coffin was not nailed shut on these other theories. A detour through Asia will lead us to the principal one among them, that of a period of cataclysmic volcanism.

3

From the roof of the world to the Deccan Traps

It was by tracking a collision not between a meteorite and the Earth but between two continents that we would flush out a new suspect that may well have caused the "Cenozoic massacres."

Early in 1980, a certain ebullience reigned at the Institut de physique du globe de Paris. Under newly signed agreements between Chinese and French research organizations, several teams would be allowed to study the Tibetan Plateau in the field. Off-limits to geologists (apart from Chinese ones, of course) for decades, the "roof of the world" was to many an object of wonder and inquiry. As long ago as the 1920s, the Swiss geologist Emile Argand had viewed this region as the result of a collision between the continental masses of India and Asia. He held that such reliefs could have arisen only by the crumpling together of what had once been several hundred kilometers of these two great continental assemblies. Heretical in a world of rigid uniformitarianism, where no one believed the Earth's crust could have undergone such major horizontal deformations, these ideas would not crop up again until more than three decades later, with the pioneering work of the British geophysicist Keith Runcorn and his colleagues at Newcastle.

The birth of plate tectonics (the modern version of Wegener's theory of continental drift) is often dated to the mid-1960s. But it was a good ten years earlier that the young Runcorn, a brilliant student of P. M. S. Blackett, had the idea of using a highly sensitive magnetometer, developed under his mentor's direction,[1] to measure the magnetization of rocks in the British Isles, and later in India; as

---

[1] Although it takes us rather far afield from the subject of this book, I cannot resist telling the reader that Blackett conceived this wonderful magnetometer, of a type called "astatic," to measure the magnetic field of a rotating copper sphere. Blackett believed that any rotating object will generate a magnetic field, for example as an electron does on the submicroscopic scale. The refutation of this hypothesis, published under the title *Results of a negative experiment*, was an important moment in the history of geomagnetism, and indeed of physics itself.

we will see, he subsequently deduced that India had drifted for thousands of kilometers since the Cretaceous Period. Runcorn was among the first to realize that the Earth's mantle is the seat of powerful convection currents, of which continental drift is only the surface expression. In the mid-1960s, the systematic exploration of ocean floors would confirm his ideas and give birth to plate tectonics.[2]

India and Asia collide

In the mid-1970s Paul Tapponnier, a young French assistant at the University of Montpellier, had joined a young American professor of geophysics at M.I.T., Peter Molnar, to study the great earthquakes that shake Asia now and then. These earthquakes occur along great faults and leave their signs on the surface for some time.[3] Paul Tapponnier had the idea of matching the maps of the epicenters of the great earthquakes against the remarkable photographs that the American Landsat satellite was beginning to take. Each photograph provides a homogeneous view of the earth's surface over thousands of square kilometers. By careful observation, Tapponnier discovered the largest strike–slip faults in the world,[4] even more imposing than the San Andreas Fault in California or the North Anatolian Fault in Turkey. Clear as a knife-cut, they can be traced in an orderly fashion, sometimes for thousands of kilometers. Two such faults partially define the boundaries of the Tibetan Plateau, primarily to the north but also to the south. So far as can be determined within the limitations of seismologic observation, it is along these great faults that the great earthquakes seem to have occurred. Moreover, these lines along which the Earth has ruptured did not come about haphazardly. To an engineer, as Tapponnier originally was, they were

2 See, for example, Note 12 in Chapter 1.
3 These are either recent traces marked in the soil morphology and still preserved from erosion or a differential erosion of formations of different natures, which the fault has brought into juxtaposition.
4 To simplify somewhat, we can distinguish three main types of fault, categorized by the movement (or deformation) of the formations they separate. "Normal" faults, such as those in the Basin and Range province of western North America, are caused by tensile forces. "Reverse" or "thrust" faults, which, for example, are abundant in the Alps, are caused by compression. Finally, "strike–slip" faults are caused by lateral, horizontal sliding. Large strike–slip faults that bound neighboring plates and transfer motion from one type of boundary (for instance a ridge) to another (for instance a trench) are called "transform faults".

curiously reminiscent of the slip lines caused in plastic soils by the weight of a building or dam.

In 1975, Tapponnier and Molnar published their conclusions in *Science*. Most of the great Asian earthquakes, they said, occur along a set of faults created by the collision of Asia with India, which began some 50 million years ago and continues today. Long successions of earthquakes have been the result. Those along the great thrust faults caused the rise of the Himalayas and the creation of the Tibetan Plateau; those along the great transform faults, hundreds and even thousands of kilometers long, favored a lateral slippage of the continental masses and the extrusion of Indochina, China, and Tibet toward the east. The findings provided both reinforcement and further details in support of Argand and Runcorn's theories.

Both the earthquake record and the satellite photographs were observations taken at a distance, hundreds of kilometers away from the "corpus delicti." Someone had to go into the field. And so after laborious negotiations Guy Aubert, Director of the Institut national d'astronomie et de géophysique, and Claude Allègre, Director of the Institut de physique du globe de Paris, signed a cooperation agreement with Chen Yuqi and Li Tingdong, officials from the Chinese Ministry of Geology. For more than three months of expeditions each year during three years, from 1980 to 1982, dozens of French geologists, geochemists, and geophysicists would have the good fortune to be the first Westerners to sample and analyze on-site, by the most modern techniques of geoscience, rocks that bore witness to the greatest continental collision Earth has known in 200 million years.

On the roof of the world, in 1981, Claude Allègre, Paul Tapponnier, and I sat up at night talking over the great debates in which our disciplines were absorbed at the time. I remember we mentioned the end of the dinosaurs, and the asteroid theory, that the Alvarezes had published a year before. None of us had worked on the subject, but we were intrigued by the elegance and audacity of the idea, and – no doubt because of the standing of its authors – basically convinced of its validity. But in any case, we weren't in Tibet for that.

Drilling at the roof of the world

The part of the mission that fell to the paleomagneticians[5] was to apply their techniques to everything that would come under our hammers or that our drills brought to the surface.[6] We have already seen how the Earth's magnetic field can be recorded and preserved almost indefinitely in the memory of rocks. Based on this field's ability to reverse its poles, we have developed a reversal scale, which is a powerful tool for correlating segments of geological time. But we have not yet used the information implicit in the actual direction of residual magnetism, as measured in the laboratory. As for a star in astronomy, or the geographical coordinates – latitude and longitude – of a site, this direction is defined by two angles, one of them known to almost everyone, and the other perhaps less familiar (Fig. 3.1). The first, declination, is the angle formed between the direction of the magnetic north (we have already seen that the blue needle of the compass actually points to the magnetic south, but never mind; the usage is too well established by now for us to correct the term) and the geographic north. The second is inclination, or dip, the angle that the magnetic direction forms with the local horizontal plane. The mean magnetic field of the Earth comes quite close to that of a magnet with two poles (a dipole) set at the Earth's center and aligned with its axis of rotation. You might remember the high-school or college experiment that involves laying a piece of paper flat on a magnet and then dusting it with iron filings, which arrange themselves into two bands of simple curves, demonstrating the "lines of force" of the dipole's magnetic field (Fig. 3.2).

Late in the sixteenth century, William Gilbert, physician to the Queen of England and a physicist in his spare time, commissioned a small sphere or "terrella" to be sculpted from a block of lodestone.[7] He had found that inclination, indicated by tiny needles left free to orient themselves close to the surface, varied regularly from

5 Our Chinese colleagues plus José Achache and Jean Besse – both of them my graduate students at that time – Jean-Pierre Pozzi, Michel Westphal (of the Institut de physique du globe de Strasbourg), and myself. Other specialists in Earth sciences sometimes call us "paleomagicians." Maybe because our findings inspire them with such awe . . .

6 Paleomagneticians often drill oriented "cores" as specimens from rocks, 2.5 cm in diameter, using a diamond-headed drill.

7 The name then given to the iron oxide now known as magnetite.

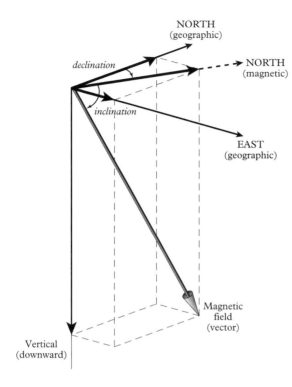

Figure 3.1
The components
of the magnetic
vector: north, east,
and vertical, and
the two angles
declination (given by
the compass) and
inclination.

the model's two poles, where these needles stood perpendicular to
the surface, to the equator, where they were tangent to the surface
(Fig. 3.2). He thus predicted, before anyone had observed it directly,
that inclination would vary continuously from vertical at the Earth's
poles to horizontal at the Equator. The globe was nothing but a big
magnet: "Magnus magnes ipse est globus terrestris"[8] is the title of
one chapter of his *De Magnete*, a treatise on magnets published in
1600 – which may be the first work of modern experimental physics.
In homage to his contribution, paleomagneticians have given his
name[9] to the fourth period of magnetic polarity (primarily reversed);
it is preceded by those of Brunhes, Matuyama, and Gauss (see Fig.
2.3, p. 30). The transient effects of the "secular change" in the
Earth's magnetic field are eliminated if you calculate the mean of
this field over several millennia: the field becomes quite similar to

8 See Note 4 in Chapter 2.
9 Just as oceanographers draw on mythology or
the memory of their great predecessors to bap-

tize underwater mountains, or planetologists do
so to baptize surface features of other planets.

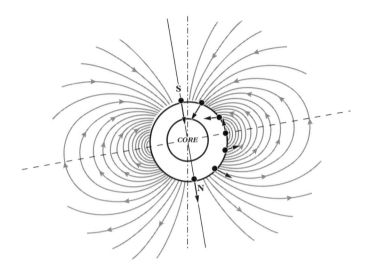

Figure 3.2
The field lines of a magnetic dipole placed at the center of the Earth. The magnetic field at the surface is indicated by a black arrow at various locations from North (N) to South (S) magnetic poles.

that of a dipole, a magnetized bar set at the center of the earth and aligned with its axis of rotation.[10]

The magnetization fossilized in rocks simply records the lines of force of the field that prevailed at the time the rocks were laid down. Declination points to where the needle of the compass would have pointed at the time and gives the direction of the magnetic north, which we will assume is identical to geographical north. Inclination, however, gives the latitude. If they can be sure their specimens haven't been altered or deformed too much, paleomagneticians can find the latitude and orientation of the continental mass from which the specimen was taken. The longitude, by comparison, is indeterminate because of the cylindrical symmetry (about the Earth's rotation axis) of the axial dipole and the field it creates.

And so, from the magnificent red sandstone of the so-called

10 At Paris, 48° N latitude, the expected inclination is 61°. Because of secular change (and the nondipolar part of the field), today it is in fact on the order of 64°. The formula that links inclination I to colatitude θ (the angular distance between the pole and the site) is quite simple and we give it as intellectual fare for the interested reader. If tan is the trigonometric function known as the tangent, $\tan(I) \times \tan(\theta) = -2$. So you can easily deduce the latitude from the value of the inclination. This is the basis for paleogeographic reconstructions of paleomagnetism, one of the first applications of which goes back to Runcorn (see the beginning of this chapter).

Takena formation, some 100 million years old, which outcrops not far from Lhasa and across a great part of the high Tibetan plateau, we were able to extract the secret of the rock's latitude at its time of deposition: about 15°N, in a tropical environment of great river deltas, more than 1500 km south of the present position. The red sandstones were subsequently folded, undoubtedly in the first phases of the collision; then they were eroded and covered again, discontinuously, by dark lavas, andesites typical of the sinking of one tectonic plate (here India) under another (the Asian part of Tibet). These lavas, dating from some 50 million years ago, show essentially no change in latitude: the southern margin of Asia, oriented more or less east and west, had hardly moved for 50 million years before the collision really took over.

So we began to get a good idea of the geography of Tibet in the Cretaceous Period and at the start of the Cenozoic. To determine the precise history of the collision and decode the lengthwise compression that occurred within the Himalayan chain, all we would need was the same information from the other side of the mountains – in other words, India. But a quick survey of the literature showed that, in fact, there were few recent data of good quality on the stable, undeformed part of the subcontinent.

And yet it was in India, as we have seen, that the modern history of paleomagnetism began, in the 1950s. Keith Runcorn and his colleagues had indeed demonstrated the drift of India, but their data, old and rather scant, did not offer the necessary precision to measure what kinds of distance had been compressed to form the Himalayas and Tibet, and to track the evolution of this compression as the collision went on.

The Deccan Traps

We had just determined the position of Asia between 100 and 50 million years ago; all we needed now was to do the same for India. The Cretaceous Period that interested us is conventionally shown in green on geological maps. The map of India displayed an immense green spot, about the size of France, or halfway between that of Texas and California: the Deccan Traps (shown in black in Fig. 3.3). This is a stack of flows of basaltic lavas, often greatly altered, most

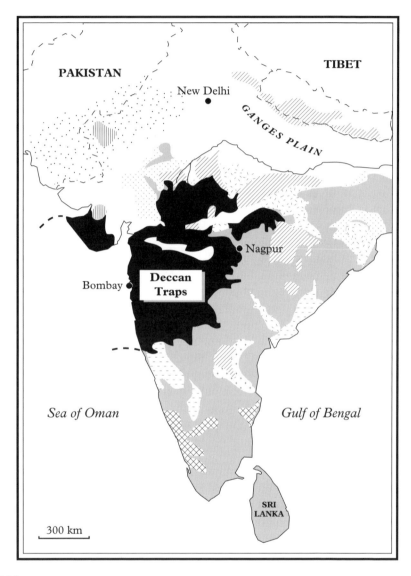

Figure 3.3
A simplified geological map of India. The Deccan Traps (flood basalts) are shown in black.
Older rocks are in gray shades, younger ones in white.

of the time forming a plateau with very low relief and covered with
vegetation. Tropical alteration has often transformed the ground into
laterite, and outcroppings of good-quality "fresh" rock are not abundant. The plateau gently slopes to the east, and the relief becomes
more marked along the West Coast of India. Rivers have cut canyons

Figure 3.4
A view of the Deccan Traps in the western Ghats, Mahabaleshwar (India), showing their stair-case-like erosion profiles (photograph by the author).

that reveal a section of the volcanic stack over a thickness of more than 1500 m. Here erosion has sculpted this relief into steps, from which the formation received its name of "traps"[11] (Fig. 3.4). These lavas had been dated by potassium–argon radiochronology at ages ranging from 30 to 80 million years ago – largely corresponding to the period for which we had data from Tibet.

Paleontologists Jean-Jacques Jaeger and Eric Buffetaut, then at the University of Paris VI, had begun a vast program of coopera-tion with several countries in southern and southeast Asia, and espe-cially with Ashok Sahni and the University of Chandigarh in India. With these colleagues, our team (Jean Besse, myself, and a new stu-dent, Didier Vandamme, who had been assigned this topic for his thesis) began sampling the Deccan basalts in 1984 and 1985, along the longest possible profiles, from Bombay in the west to Nagpur and Jabalpur in the northeast. A few months after our return to the laboratory, we were confronted with a result we hadn't expected at

11 The root of the word appears in several Nordic languages, with the sense of "stair steps." The term was introduced by Bergman, a Swede, in 1746.

all: most of the specimens had the same polarity – in this case reversed. Where we had sequences of superimposed lava flows, we almost never found a reversal. By compiling and critically analyzing all the results published since the 1960s, we soon found that although hundreds of meters of superimposed lava had been sampled, we never observed more than two reversals, and more often only one or even none. One thing quickly became obvious: the stack of lavas in the Deccan, cumulatively more than 2000 m thick, had recorded only two reversals of the Earth's magnetic field! The 1985 expedition was dedicated to testing what was not yet even a hypothesis in a still little-sampled region of the Deccan. The result was positive. All our observations were only apparently complex; they could be explained if we assumed that the lava flows in the traps formed vast but very slight undulations, which in fact were practically invisible to the naked eye and yet were entirely compatible with volcanologists' ideas of how they were put in place.

Five hundred thousand years or fifty million years?

If we remember that the Earth's magnetic field underwent dozens of reversals over the period from 80 to 30 million years ago – ages suggested by radiochronology – the paleomagnetic result we had obtained in June 1985 was unexpected, to say the very least. According to us paleomagneticians, this volcanism could not have lasted more than a few million years at the outside. Where was the error? First we had to re-examine the "absolute" ages provided by potassium–argon dating. Using this same method at the Institut de physique du globe de Strasbourg, Raymond Montigny had obtained new ages. He pointed out that some of the spread in the results might have been caused by the degree of alteration of the basalts. Using the latest argon isotope technique (called "argon 39–40"), Gilbert Féraud in Nice narrowed the spread to only 4 or 5 million years, at between 63 and 67 million years ago, or a duration ten times less than had been indicated by the first potassium–argon dating. We began to entertain serious doubts about the reliability of the old ages.

For their part, our paleontologist colleagues had found the remains of dinosaurs and even mammal teeth in fine sedimentary strata in the Deccan, vestiges of lakes that had developed temporarily between eruptions. So volcanism certainly began before the end of the Mesozoic. Here I cannot resist retelling the unbowdlerized tale of one of our paleontologist friends' discoveries. Near Nagpur, one of Ashok Sahni's thesis students had been digging for years, filtering and analyzing tons of sediment with no great success. While we were there, drilling a few cores to look at paleomagnetism, another paleontologist, Henri Cappetta, stepped aside to answer nature's call. The resulting less-than-classic agent of erosion revealed, under the very eyes of the unhappy student, a little white fragment a few millimeters long, which immediately attracted attention. Within minutes, and to my great admiration, Henri Cappetta had identified a tooth of a freshwater ray, until then unknown outside the sediments of the last part of the Cretaceous Period (the Maastrichtian) in Niger! So the volcanism had begun during the last stage of the Cretaceous.

Jean-Jacques Jaeger then directed our attention to the results from a core drilled off Bombay. Several volcanic flows – undoubtedly distant evidence of the traps – had been cut through, sandwiched between strata of marine sediments, which are easier to date and correlate against the global scale than are the continental sediments on which the traps generally rest in India. Under the first flow lay a zone of planktonic foraminifera characterized by a species with the sonorous name *Abatomphalus mayaroensis:* in fact this zone was the last and thinnest subdivision of the Maastrichtian – in other words, the last million years of the Mesozoic. The message of this tiny marine fossil was far more precise than the one left by the last dinosaur bones in the continental environment.

The conjunction of paleomagnetic, geochronological and paleontologic results permitted only one correlation. The "reversed" period during which the main body of the traps had been extruded appeared to be none other than chron 29R (Fig. 3.5), precisely the one in which the Cretaceous-Tertiary boundary had been found in Gubbio! A statistical calculation allowing for the thickness and polarities of the traps, and the duration of chron 29R, led us to conclude that this enormous volcanic mass had been laid down in less than half a million years, one hundred times more rapidly than we could have imagined at the start. Moreover, with the best available temporal precision, the eruptions seemed to coincide with the biological events

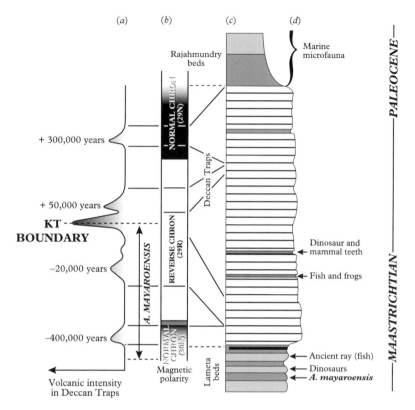

Figure 3.5
A synthesis of paleomagnetic, paleontologic, and geochronologic data from the Deccan Traps.
(a) Tentative model of changes in volcanic intensity with discrete peaks; (b) magnetic polarity
(black, normal white, reversed); (c) geological section with under- and overlying sediments, and
the lava flows sandwiched between the two; (d) some key fossils and geological ages.

that had marked the transition from the Mesozoic to the Cenozoic:
the absolute age of the traps was the same as that of the disappearance of the dinosaurs. It was hard to see this as only a coincidence, and in 1986 we proposed that volcanism might, in fact, have
been the principal agent in the mass extinction of species. Suddenly,
unwittingly, we had become actors in the debate over the KT boundary: and we were not in the "impactist" camp. We would soon discover that others had not waited for us and had proposed, on the
basis of far more tenuous data, that the end of the dinosaurs had
been caused by volcanic exhalations without parallel in modern
times. A "volcanist" movement was already forming ranks against
the "impactists."

4
___

The volcanic scenario

To some extent, we had failed to achieve our initial goal – to describe the movement of India over tens of millions of years. And this was simply because instead of covering the long period in which we had originally been interested, the rocks we studied represented only a brief instant of geological time. On the other hand, we had unexpectedly discovered that the Deccan Traps might be the cause of the extinctions at the KT boundary. Might this exciting result, which we were preparing to publish late in 1985,[1] already have been found by someone else? Could this idea, which we ourselves found a little revolutionary (and which would certainly be unorthodox in the eyes of many specialists in continental basalts), have been formulated before, as often happens in any science? This had already been the case for the Alvarezes' asteroid, which was suggested several times before their work, for example by the great Harold Urey in 1973. But the Alvarezes' contribution, with the discovery of iridium, had provided the impetus for a qualitative and quantitative leap forward.

Precursors

In December 1985, I was at the Fall meeting of the American Geophysical Union in San Francisco. This annual convention has become a world gathering place for geophysics. Though some of our results had been written up since June, none had been published. Spending one Sunday with my friend Marvin Chodorow, a Stanford professor and a former director of the Microwave Laboratory there, I brought up our ideas to him. Always alert to new developments well beyond his own discipline, he thought he might have read a paper on the subject lately in the weekly journal *Science*. And so I

1 Two articles appeared, in the *Comptes rendus de l'Académie des sciences* Paris and the international journal *Earth and Planetary Science Letters*.

discovered the article in which geophysicists Chuck Officer and Chuck Drake, of Dartmouth College, attacked the impact hypothesis and argued that the events at the KT boundary might have lasted at least 10,000 to 100,000 years. According to them, the physicochemical anomalies at the boundary might be explained just as well by an Earthly source. They even noted that the Deccan Traps had undoubtedly been produced quite rapidly (though they could not say just how rapidly), and at the right period. As any researcher will easily understand, reading this article filled me with two conflicting emotions: bitter dismay that we were not the first to formulate the scenario and excitement that our ideas were supported by independent work.

Later, I found a half-dozen articles that stated the premises, or even provided a well-constructed presentation, of a substantial part of our hypotheses. Some had been published in specialized journals of paleontology or climatology that were unfamiliar territory to a geophysicist like me. But I could not have been alone in this, as the articles had been little cited since, and I had a hard time getting references or copies of some of them. From this bibliographic research I want to cite some names and dates: the names of people who now seem to have been our main precursors, and the publication dates of their first ideas.

Back in 1968, Michael McElhinny, one of the greats of world paleomagnetism, discovered the small number of reversals in the Deccan Traps and suggested, on the basis of a still-uncertain chronology of magnetic-field reversals, that the main phase of volcanism need not have lasted much more than 5 million years. Based on this result, and on the ages of the traps and of the KT boundary, which he set at between 63 and 70 million years, Peter Vogt (in 1972) appears to me to have been the first to link major volcanism with extinctions. Estimating the duration of these catastrophes at some 5 million years, he lumped them together with other volcanic phenomena such as the birth of the North Atlantic Tertiary Volcanic Province,[2] which we now know to be appreciably younger. According to him, the extinction mechanism was the injection of vast quantities of toxic metals

---

2 This is one of the names given to the major assembly of volcanic rocks that outcrops in the northwestern British Isles (including Mull, Rhum, and Skye) and all along the eastern coast of Greenland.

into the atmosphere, whereas in normal times the concentration of these in sea water is infinitesimal. We should also mention the great American geophysicist from Princeton, Jason Morgan, principal among the fathers of plate tectonics;[3] he too proposed (in 1982) that the KT extinctions should be associated with the great Deccan Traps.

Carbon dioxide and the biological pump

The first detailed model of the possible climatic effects of the Deccan Traps was proposed back in 1981 and later defended through thick and thin in the midst of a climate of astonishing hostility (partly from Luis Alvarez) by Dewey McLean. A specialist in the carbon cycle in the atmosphere and hydrosphere, McLean was persuaded on the grounds of far more tenuous data than our own that the extinctions coincided with the Deccan volcanism. He emphasized that the traps must have injected ten times more carbon dioxide into the atmosphere than the atmosphere contains today. The oceans, warmer on average than they are now, would not have been able to dissolve this gas, which, therefore, would have accumulated in surface waters and the atmosphere. The resulting chemical and physical conditions would have considerably reduced photosynthesis and the production of calcium carbonate, resulting in a "dead ocean" and explaining the presence of the stratum of clay in many sections at the KT boundary.

Under normal conditions, single-cell algae extract carbon dioxide from the air and water and use it to build their calcareous skeletons. When they die, their solid remains fall to the bottom and are incorporated into limestone sediments. This is what we call the "biological pump," or the Williams–Riley pump. So most of the Earth's carbon is stored, not in water or the atmosphere in the form of carbon dioxide, but in rock, such as the rocks that form the famous White Cliffs of Dover. McLean proposed that the acidity and composition of surface waters, profoundly modified by the dissolution of a portion of the volcanic emissions, might have killed off this algae and shut down the pump. If it were suddenly shut down, the carbon dioxide content of the atmosphere would double in 25 years! In fact

3 Along with Briton Dan McKenzie and Frenchman Xavier Le Pichon.

the pump absorbs 2×10^{15} moles of carbon each year, a thousand times more than is produced by all the active volcanoes in the (modern) world. If McLean's style seems rather daringly self-assured in view of the data he had at the time, this no doubt partly reflects the aggressive atmosphere in which he was trying to defend his ideas.

Marc Javoy and Gil Michard, of the Institut de Physique du Globe de Paris and the University of Paris VII, have since modeled the climatic effect that would have resulted from some 50 volcanic events on the scale of those that created the great lava flows in the traps, on the hypothesis that each of them would have lasted several years and had been separated from the next by several millennia.[4] According to them, the effect of each injection of carbon dioxide, amplified by the gradual destruction of the biological pump and above all the repetition of these injections, which the ocean would be less and less able to absorb, ultimately multiplied the atmosphere's carbon dioxide content by a factor perhaps of more than five. Through the greenhouse effect, this would increase the temperature of the lower atmosphere by more than $10°C$ – a considerable rise.[5] These models involve great uncertainties, and other researchers envision thermal effects of only a few degrees. However, the likely short-term consequences of great volcanic eruptions are a terrible winter and acid rains, as the recent study of several great historic eruptions allows us to assume.

Ben Franklin's idea

Benjamin Franklin was the first to suggest that volcanism might have significant effects on climate. As the young American republic's ambassador to Paris in 1783, he was struck by widespread evidence of an exceptional change in the climate of Northern Europe: the amazing color of the sky during the day and at sunset, a strange permanent fog, abnormal heat in the summer. A thick bluish haze drifted across Europe and was reported 50 days later in China, in the Altai Mountains. The winter of 1783–84 was particularly harsh in Europe.

4 Their results, published as a rather obscure abstract, have not received the recognition they deserve.
5 A modification of just half a degree in the mean temperature of the lower atmosphere is already a significant event. We will return to this point below.

Having heard of volcanic eruptions in Iceland, Franklin had the idea of linking the two phenomena. In a communication to the Manchester Literary and Philosophical Society,[6] he argued that the fog was caused by an eruption, and that it interfered with the passage of sunlight. The measurements taken at the time confirm that temperatures in the Northern Hemisphere were the lowest in more than two centuries. The effects began in the summer of 1783 and continued for at least two years. The great Icelandic volcanologist Thorarinsson has reconstructed the events that took place in Iceland from June 1783 to March 1784. In the southeastern part of the island, near Mount Laki, a series of earthquakes were the prelude to the opening of a fissure. Explosion craters developed across a distance of more than 25 km. Within just a few months, the fissure had extruded about 12 km^3 of basaltic lava, the largest eruption of this type in historical times. The emitted gases, chiefly dioxides of carbon and sulfur, destroyed vegetation, grassland, and crops and led to the greatest famine the island had ever known: between 50 and 80 percent of the livestock perished, along with a quarter of the human population. Similar events, though less marked, followed on the great eruptions of Tambora in 1815 and Krakatoa in 1883.

Great eruptions, particularly those of volcanoes with viscous, acidic lava, produce great quantities of ash. Injected into the atmosphere, this ash reflects some sunlight and may cause a drop in temperature. Until the 1950s, such dust was blamed for most of the climatic effects of volcanic eruptions. It was the 1963 eruption of the volcano Agung that attracted attention to the probable role of sulfur in the expelled aerosols. Escaping as sulfur dioxide, it combines with water to form tiny droplets of sulfuric acid. In sufficient quantity and dispersed throughout the upper atmosphere, these droplets may cause cooling, a partial destruction of the ozone layer, and, finally, acid rain. They remain in the atmosphere far longer than dust, which generally aggregates and falls rather quickly, no more than a few hundred kilometers away from the eruption site.

6 These conjectures are reproduced in H. Sigurdsson's excellent article in *EOS Transactions* *of the American Geophysical Union*, Vol. 63, AGU Washington, 1982.

Volcanism and climate

By now it is well established that the climatic effects of an eruption depend not only on the total mass of ash and gas the volcano emits but also on the composition of these gases, the rate at which they are ejected, and the height to which the eruption column rises in the atmosphere. Of course, they also depend on the volcano's latitude and on how efficiently atmospheric air currents disperse the aerosols the eruption produces. An underwater eruption has negligible effects compared with an eruption in the open air. Volcanoes in island arcs, those regions where one tectonic plate plunges under another to return to the earth's mantle, generally produce lavas richer in silica, and their eruptions are particularly explosive. The recent eruptions of Mount Saint Helens in 1980, El Chichón in 1982, Nevado del Ruiz in 1985, and Pinatubo in 1991 were of this type. They may emit tens of cubic kilometers of material within a short time, and in the case of the celebrated "Plinian" eruptions (such as the eruption of Vesuvius in 79 AD that cost Pliny the Elder his life) their columns may rise more than 10 and sometimes as much as 50 km into the air. In these eruptions, ejection rates may reach a billion kilograms per second. The eruption of Tambora, in 1815, produced 50 km^3 of lava; but that of Toba in Indonesia, dated to 75,000 years ago, produced several *thousand* cubic kilometers of lava and spread ash as far as 2,500 km away.[7]

Recently, a very precise physicochemical approach to these remarkable events has become possible by studying not only the products ejected in these great eruptions but also, unexpectedly, ice cores sampled in Greenland and the Antarctic. The strata of snow that build up on the icecaps and turn into ice imprison bubbles of gas. The chemistry of these gases and of the now-frozen water preserves a record of the climate. And thus it has been possible to show that the determining factor in the climatic impact of an eruption is indeed its sulfur content (Fig. 4.1). The eruption of Mount Saint Helens, very low in sulfur, had an effect of no more than one-tenth of a degree Celsius. The sulfur-rich eruptions of El Chichón and Pinatubo had

---

7 The thickness of this ash after compaction amounts to 0.1 mm at this distance. Despite appearances, this implies a very large total volume, as the reader can easily confirm by a simple geometric calculation.

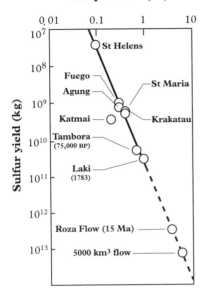

Decrease of mean atmospheric temperature (°C)

Figure 4.1
Decrease in mean atmospheric temperature following massive injection of sulfur by a volcanic eruption, as a function of sulfur input. Extrapolation from actual historical eruptions to the much larger flood basalts is tentative (dashed). (After K. McCartney.)

more marked effects, on the order of two or three tenths of a degree. The largest historical eruption of an island-arc volcano, that of Tambora, released several tens of billions of kilograms of sulfur,[8] lowered the temperature in the Northern Hemisphere by 0.7°C, and destroyed nearly a tenth of the ozone layer. If we extrapolate the resulting correlation between sulfur production and lowered temperature, the effect of the gigantic Toba eruption must have been 4°C.[9]

From Hawaii to the Deccan: effects of scale

Hot-spot volcanoes, which we will return to below, are another type. The volcanoes of Iceland, Kilauea in Hawaii, and Piton de la Fournaise on Réunion are among the best known. Emitting fluid basaltic lava, generally they do not erupt explosively or produce great

---

8 Each year our civilization emits 50 billion kg of sulfur into the troposphere.

9 But it is not certain that we are entitled to do

this, since nonlinear effects may very well take over in these paroxysmal events.

quantities of dust. But they are often much richer in sulfur than the island-arc volcanoes (which have concentrations of several parts per thousand). For example, the eruption of Ben Franklin's volcano, Laki, injected tens of billions of kilograms of sulfur into the atmosphere – more than Tambora did, even though the latter emitted five times more lava.

Yet the Laki Flow is nothing compared with the huge flows that erupted when the great traps were laid down. A single flow in the Columbia River Traps (dating from 16 million years ago, in the Miocene Epoch), the great Roza Flow, amounted to 400 km³. Its estimated effect on the climate was 4°C. The big problem is to work out whether these nonexplosive volcanoes are able to project the aerosols they produce beyond the tropopause, into the stratosphere. If not, the erupted matter would only have a local effect and would not be dispersed over the globe.

Richard Stothers, Steve Self, and several of their colleagues at the Hawaii Center for Volcanology have calculated the altitude of the atmospheric plumes that form by convection above the great fountains of lava typical of fissural basaltic volcanoes. These calculations show that the great plumes emitted in Hawaii climb to altitudes of "only" 6 km and, therefore, remain in the troposphere. Those of Laki must have risen to 10 km and grazed the tropopause. The Roza Flow, with its 100 km long fissure and an eruption rate of 1 km³ of lava from each kilometer of fissure each day, lasted for several days and up to several months. It must have produced a fountain of lava nearly 1000 m high, and a plume that easily reached the stratosphere. In the Deccan Traps, the fissures must have been more than 400 km long, and the volume of some flows must have reached several thousand cubic kilometers. So we can imagine the effect of a single eruption of this type, the like of which the Earth has not seen in millions of years: a darkened atmosphere (sunlight would have been reduced by a factor of a million, leaving it no brighter than a night with a full moon), abnormal climatic effects for from one to ten years, with a temperature drop of nearly 10°C and abundant rains of sulfuric acid. The volcanic winter ensuing upon each major eruption would have been followed by the longer-term consequences of an injection of carbon dioxide and the resulting greenhouse effect, as we have seen above. We can also imagine what would have happened if many such events were repeated over several tens or hundreds of thousands of years.

Volcanism and iridium

The age and duration of the laying down of the Deccan Traps, and the considerable climatic disturbances that must have resulted, suggest a plausible scenario of an ecological catastrophe likely to result in the massive extinctions at the KT boundary. This scenario is furthermore quite close to the one proposed by the asteroid advocates – except that its duration is counted in tens or hundreds of thousands of years, not days or months – and just as terrifying. But is the volcanism of the traps able to account for the physical and chemical anomalies on which the impact scenario was founded? Walter Alvarez was still answering this question with a categorical "no" on all counts in 1986. However, the work of Chuck Officer and Chuck Drake, and above all the article I had just read in *Science* in 1985, followed by a series of other works published afterward by large numbers of researchers, would dampen at least some of this certainty. For rigor's sake, we must now review all the arguments and observations.

First of all, what about the famous stratum of clay at the KT boundary, interpreted as the product from the alteration of the dust very briefly deposited all over the globe after the impact? In fact, this clay is not present everywhere. At Stevens Klint or Gubbio one can observe large numbers of interbedded strata of identical clay, not only at the boundary but also several meters above and below it. In Denmark, the stratum consists primarily of a mineral of the clay family, a smectite rich in magnesium. The presence of a characteristic feldspar,[10] labradorite (an attractive iridescent blue mineral much favored by banks and cemeteries for their decor), and even the composition of the clay itself suggest a product from the alteration of basaltic volcanic ash.[11]

---

10 Feldspars are silicate minerals abundant in most magmatic rocks, accompanied by quartz, micas, amphiboles, pyroxenes, and olivines. The proportion of these major components is the basis for the classification of these rocks. The calcium, sodium, or potassium content of a feldspar, in turn, is a more refined tool for classification that provides access to the chemical and thermodynamic conditions under which the rock containing them was formed.

11 This is the view of Chuck Officer, among others.

Second, there is the presence of iridium. Its concentration in the sections at the KT boundary varies widely, from 0.1 to 100 p.p.b. Here we may wonder just what the definition of an iridium "anomaly" is. Enrichments of from one to several parts per billion are known in rocks of undisputedly terrestrial nature. The maximum amount of iridium has been found at the base of a coal vein in North America. In Europe and New Zealand, the element is associated with a stratum rich in organic matter. Now, the marshy environment that favors the deposition of coal is capable of concentrating metals such as iridium. Hansen discovered a correlation between the distributions of iridium and carbon, and that fossil bryozoans (generally marine animals that live in colonies) were encrusted with this carbon. He argued that this carbon, deposited in successive episodes over a rather long total duration, was the carrier of iridium.

Then it was discovered that elevated iridium concentrations were not limited to the strata from the precise time boundary between the Cretaceous and the Tertiary Periods but generally extended for several meters on either side. Robert Rocchia described numerous sections in which he and other researchers saw proof that the source of iridium, whatever it may have been, had been in operation not just for a few years, but for several hundred thousand years.[12] For a long time, supporters of the asteroid theory simply denied the possibility that volcanism might emit large quantities of iridium. But back in 1983 some Americans, Ed Zoller and his colleagues, very significantly showed iridium in the aerosols emitted by Kilauea in Hawaii. A short time later, two Frenchmen, Toutain and Meyer, discovered it in the fumaroles of Piton de la Fournaise on Réunion. Hot-spot volcanoes, whose origins many think lie deep in the upper mantle and not in the Earth's crust, are, therefore, capable of producing iridium.[13]

At least some geochemical indicators show that the products accumulated at the KT boundary could not have come from the Earth's crust alone. But several of these products might just as well have come from the upper mantle as from an extraterrestrial body. The

12 Rocchia now thinks the iridium was in fact diffused chemically rather far from the stratigraphic boundary.
13 In marine cores, layers of volcanic ash from recent Arctic eruptions have been discovered with iridium concentrations of up to 7 p.p.b. Researchers have lately observed this element in the volcanoes of Kamchatka. But these two cases do not involve hot-spot volcanoes.

signatures of these two types of "geochemical reservoir" are extremely similar. Our indirect knowledge of the chemistry of the deep parts of the globe is based largely on our direct knowledge of meteorites, many of which are thought to be only slightly altered evidence of the first phases of the state of matter at the time when the Earth and the other planets were formed.[14] The ratios of osmium isotopes, for example, which Luck and Turekian said might be characteristic of the mantle, are not all that far from those that Luck himself, with Allègre, observed in the terrestrial rocks of the great Bushveld complex in Africa.

Could the total quantity of iridium deposited on Earth at the KT boundary, in the order of 200,000 or 300,000 metric tons, have been produced by the Deccan Traps alone? Most experts say no. However, volcanism might do a better job than an impact in providing an explanation for the observed levels of arsenic, antimony, and selenium, which are fairly scarce in meteorites.

Volcanism and other anomalies of the KT boundary

The third element: the microscopic spherules of heavily altered basalt. As we saw, the meteorite partisans interpreted these as molten droplets formed by the impact and subsequently dispersed. Some of these spherules are hollow; others are not. Some have been identified as the remains of microscopic fossil green algae, replaced by secondary minerals. Others consist of a feldspar, characteristic of the alteration of what must have originally been droplets of molten basalt. Still others have even been identified as the organic eggshells of present-day insects, included by mistake with the mineral samples in the field! Tiny spheres of amorphous glass found in the sections at the KT boundary in Haiti have been reinterpreted by backers of the impact hypothesis as volcanic droplets, on the basis of petrological and geochemical features. Similar spheres have apparently been found in the strata of tuffs from the Paleocene Epoch, undoubtedly of volcanic origin, in western Greenland. These spheres are rich in iridium and contain inclusions of magnetite high in nickel, very similar to those Rocchia and his colleagues think can only be associated

14 Chapter 2, Note 17.

with a meteorite.[15] Do microspherules really make it possible to distinguish between impact droplets and volcanic droplets?

One of the strongest arguments in favor of the impact theory does indeed seem to be the presence of the shocked quartz as discovered by Bohor. Indisputably observed after exceptionally brief, violent shocks, whether natural or artificial (a meteorite, a nuclear explosion, or laboratory experiments), shocked quartz has never been seen in the products of volcanic eruption. Some calculations nevertheless suggest that an explosion such as the one that occurred at Mount Saint Helens could have caused such crystal defects. But this model has come in for criticism and remains unsupported by any observation from among the ejecta of Mount Saint Helens. Neville Carter and Chuck Officer thought they had found characteristic shock features in grains resulting from the colossal eruption of Toba, but Jean-Claude Doukhan could not find them by analysis under the transmission electron microscope. Shocked quartz has been observed in enigmatic geological objects called "cryptoexplosion features," such as the great Vredefort complex in Africa. But specialists are just as divided on the origin of these objects: impact, or volcanism from the mantle?

Diamonds under the Ghats?

It's important to remember that the Deccan volcanism has had no equivalent for millions of years, and that even the Toba eruption is minuscule compared with what must have happened back then. The basaltic, fluid, "calm" volcanism of the Deccan must have been preceded by substantial periods of explosion, as the magma rose to the surface and digested the acidic rocks of the old continental crust of India. Such explosive phases produce eruptions of kimberlites, those astonishing lavas that enclose diamonds formed in the mantle.[16] Marc Javoy and I have argued that the global geochemical anomalies in strontium, recorded in the marine environment at the KT boundary (and measured in the shells of organisms that lived then), can be interpreted as the trace of such an explosive, acidic volcanism, which

15 But these results have only been published as a brief abstract, and Robert Rocchia, who has had several specimens in his hands, thinks these spinels have nothing to do with cosmic spinels.

16 See the article by V. Sauter and P. Gillet, Les Diamants, messagers des profondeurs de la Terre, La Recherche, 25, 1238–1245, 1994.

was then covered over by basaltic flows and has not yet been uncov-
ered by erosion. Could the eruption of the Deccan Traps have been
preceded by explosive kimberlite eruptions of exceptional violence,
capable of producing the defects observed in the grains of quartz and
zircon torn from the walls of the eruption channels? Might diamond
mines lie still unsuspected, a few kilometers beneath the plateaus of
the Western Ghats? The question has regained momentum with the
claimed discovery by the Danes Hansen and Toft of quartz grains
presenting multiple families of crystal defects in a layer of ash–ash
resulting from an explosive acidic volcanism in Greenland that dates
to the Upper Paleocene Epoch. If this observation is confirmed, we
would then have proof that these defects can also have been produced by explosive volcanism. In the future, we should look for ash
deposits far from craters and close to known outcroppings of kimberlites, where we might find quartz that was shocked and not
reheated. But such outcroppings are very rare, because most of the
time they have been worn away by erosion. We must perforce
acknowledge, on the one hand that direct evidence of the production of shocked quartz by volcanic eruptions is almost nonexistent,
and that, on the other hand, the arguments that present them as the
best element for diagnosing an impact are indeed very strong.

Traps and extinctions: a lasting catastrophe

The Deccan volcanism at the end of the Cretaceous Period did not
occur in isolation. Major geochemical and mineralogical changes
have been found in cores taken from the Walvis Ridge in the South
Atlantic, indicating an intense volcanic activity in this region over
several hundred thousand years. This duration of the anomalies
surrounding the KT boundary seems to me in itself an impressive
argument in favor of the volcanic model. As a great many observations have attested, the ecological and physicochemical crisis seems
to have started at least one or two hundred thousand years before the
level where the iridium was found, and to have continued for at least
one or two hundred thousand years afterward, with a certain number
of paroxysmal phases. This would agree with the stepwise disappearance of fossils found in several KT sections. Researchers have frequently cited the possibility that anomalies originally concentrated at

a single stratigraphic level might later be diluted and diffused over a greater thickness. Burrowing organisms can overturn the bottom on which they dwell to thicknesses of 10 cm or more. Some chemical species may be transported over longer distances by interstitial fluids. But no one has been able to see a way of "mixing" physical objects like spherules, quartz grains, and, above all, fossils over greater thicknesses.

Any scenario proposed for the disappearance of Mesozoic species must account not only for the nature of the various physical and chemical anomalies that the event left behind in the rock but also the selectivity of the disappearances and their time sequence. The hypothesis that lays the blame on the eruption of the Deccan Traps is compatible with some of the observations but not with the shocked minerals. But these tests are not unequivocal and do not eliminate the rival impact scenario. The main argument remains the very existence of the traps themselves, and the chronological framework we have established: the enormous volume of lava, the brevity of the eruptions, and the matching of dates.

An ecological disaster movie

We have already seen what climatic events might be provoked by an injection of carbon dioxide, sulfur dioxide, and hydrogen chloride gases: short-term cooling together with destruction of the ozone layer,[17] leading to an increase in ultraviolet radiation reaching the surface, acid rain, and over the longer term perhaps a greenhouse effect, with warming and intensified acidity in surface ocean waters.

At the end of the Cretaceous Period, a major and global shrinkage of the oceans had been in progress for several hundred thousand years, resulting in climates which were more and more "continental," with more and more marked seasonal variations in temperature. This marine regression might not be unrelated to the gigantic eruptions of the traps. The variations in sea level, the speed with which the ocean floor forms at the mid-ocean ridges, and global volcanism might in fact all reflect variations in activity in the upper mantle of the Earth.

17 On the problems of the ozone, see for example G. Mégie, *Ozone: l'équilibre rompu*, Paris, Presses du CNRS, 1989.

The first appearances of volcanism began in India, at 30° S lati-tude, episodically – let us say a little more than 65 million years ago. After two or three hundred thousand years, a paroxysmal phase occurred, lasting several thousand years (or at most a few tens of thousands of years). The eruptions continued, in the Deccan and elsewhere (for example the South Atlantic), and the last convulsions were over about 0.5 million years after the process began. In the sea, organisms with metabolism depending on dissolved carbonates began to suffer with the first eruptions, and increasingly severe crises coin-cided with more closely spaced or more intense volcanic episodes. The Foraminifera suffered more and sooner than the coccol-ithophores (algae armed with delicate calcareous plates, whose fos-sil remains are called coccoliths). Today, coccolithophores can withstand substantially more acidic waters, characterized by a pH[18] as low as 7, whereas Foraminifera cannot endure less than 7.6. Among species now alive, smaller forms with less ornate skeletons and species living in cooler waters at higher latitudes are better able to cope with increases in water acidity or decreases in temperature. It was precisely such Cretaceous species – small, rather plain, from high latitudes – that best withstood the several hundred thousand years the crisis lasted. And the first Tertiary organisms to appear also have these characteristics. Organisms whose skeleton consists pri-marily of silica, such as diatoms, radiolarians, and flagellates, some forms of which can even survive in polluted lakes whose acidity is considerable (pH 4!), are less affected by the crisis.

In the oceans, there is a depth below which carbonates dissolve. Carbonate skeletons deposited below this boundary are not pre-served. Under the effect of the acidification of ocean water, this "compensation depth" rises significantly, which in part explains the absence of carbonates and the dissolution phenomena observed in so many sections from the KT boundary. Species living near the coast, already affected by variations in sea level, are particularly vulnerable when the crisis reaches its climax.

Acid rain makes some freshwater lakes in the continental domain less habitable. We can understand the selective aspect of extinctions

18 Readers will recall that pH is a measure of the acidity of a solution, which will be neutral, acid, or basic depending on whether the pH is equal to, less than, or greater than 7.

when we realize that different species of fish have very different tolerance thresholds today. Trout cannot tolerate a pH of less than 5.5, while some perches survive in waters with a pH of 4.7. The plankton species that can surround themselves with a cyst when ambient conditions become too harsh are almost unaffected.

In emerged areas, the distribution of ecological stresses is extremely varied – a climate sometimes more and sometimes less continental, different temperature extremes, dust, ultraviolet radiation, or acid rain. Vegetation is subjected to these stresses, although here the paleontologic record is not very clear. I have already mentioned the "fern peak," which is interpreted as what resulted when hardier, more generalist, and opportunistic species reconquered devastated land. Eric Buffetaut thinks the breaking of the food chains based on plants and plankton is one of the keys to understanding the extinctions in the emerged domain. Dinosaurs always suffered a high rate of extinction and replacement, a sign of their capacity for evolution and adaptation. But at the time of the crisis, replacement could no longer keep pace with the disappearances. Those species that pulled through – placental mammals, birds, amphibians – undoubtedly did so for a variety of reasons, among which we may presume were simply their small size and the concomitant very large numbers of individuals, as well as their nocturnal habits, their tolerance of temperature changes, their underground habitat, and what they ate (as root eaters, carrion scavengers, eaters of organic matter in various forms, etc.). But it remains a mystery why some reptiles with a certain lifestyle survived, yet all species of dinosaurs with quite a similar lifestyle did not.

This picture of our planet's climate 65 million years ago, this poisoning of the atmosphere with volcanic gases for several hundred to thousand centuries, these mass extinctions of species that had dominated the Earth and the waters for so long, all seem to me to be compatible with many of the records that geologists, geochemists, and geophysicists have extracted since the 1960s from the archives preserved in rock. The volcanic catastrophe may at first glance seem less violent than the impact of an extraterrestrial body. Yet it is no less intense, nor are its consequences less impressive. But the duration of a human life does not offer the necessary perspective to perceive that.

5

Plumes and hot spots

If the volcanic scenario can explain the mass extinctions at the end of the Mesozoic, we must stop and dwell for a moment on the nature and geological meaning of those astonishing formations called traps. What is their structure? What is their relation to the surrounding crust on which they rest? When were they laid down, and how? Do they have any modern equivalents? Is the activity that gave birth to them now extinct, or does some trace of it still survive somewhere? Finally, how deep down were they generated? To try to answer some of these questions, we are about to embark on a voyage that will take us, perhaps not to the center of the Earth, but halfway there at least.

From the Deccan to Réunion

First of all, are the Deccan Traps an isolated structure? They cover an older continental crust, with which they have no relation and which contains nothing in either nature or composition that would plainly announce their presence. Farther north rises the great chain of the Himalayas, formed as India slowly rams against the Asian continent. The stress and deformation forces orientated south to north compress up the east–west mountain range. Could this compressive action somehow be related to the length of the gigantic, gaping fissures through which the lava surged? Apparently not. The fissures are variously oriented, some of them parallel to the Bombay coast, others undoubtedly located under the valley of the Narmada, almost perpendicular to that coast. In fact we know too little about the geometry of these orifices, covered by the traps, to be able to link them in more detail to the surrounding structures.

However, if we carefully look at the morphological map of the ocean floors southwest of the coast (Fig. 5.1), we soon notice a north–south alignment of submarine mountains, some of which emerge to form islands: from north to south, these are the archipelagoes of the

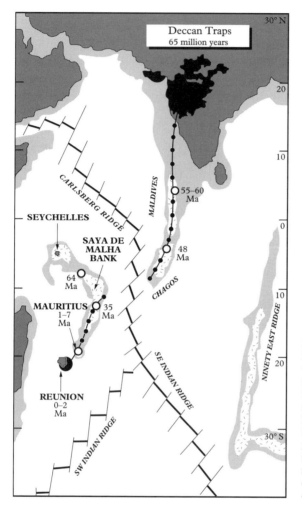

Figure 5.1
Part of the Indian Ocean, showing chains of seamounts leading from the active Réunion hotspot, to the Deccan Traps. Age progression (in Ma) is shown where actual measurements were carried out on dredged samples.

Laccadives, Maldives, and Chagos. This alignment is interrupted to make way for the "normal" ocean floors created by the Carlsberg Ridge, which runs from the Triple Junction of Rodrigues in the south toward the Gulf of Aden and the Afar region of Africa in the northwest. The triple point, which Roland Schlich of the Institut de physique du globe de Strasbourg, Philippe Patriat, and their associates have studied in very great detail, is the meeting place of three great lithospheric plates: the African, Indo-Australian, and Antarctic. Some 20 Ma ago the Indian Ridge lengthened to the northwest, then to the west, and tore apart the continental crust of Africa, thus opening up the Gulf of Aden between Arabia and Somalia. It then penetrated inland and formed the Afar depression. French teams have played a significant role in understanding the impressive phenomenon of the birth of this new ocean.[1] To the south of the Indian Ridge we find abnormally shallow reliefs, the banks of the Seychelles and Saya de Malha, and even farther south, a chain of islands that ends with Mauritius and finally Réunion.

In 1986, after dating the Deccan Traps, we wondered whether this string of islands, which still links the active volcano of Piton de la Fournaise to the extinct volcanism of India, might not represent some sort of umbilical cord, indicating a shared origin. All the more so when we discovered, thanks to paleomagnetism, that the traps had formed somewhere around 30° S latitude, not far from the current latitude of Réunion. In this discovery we had two guides, the theory of hot spots and plumes set forth by the Canadian researcher Tuzo Wilson and by Jason Morgan, and certain experimental results regarding the instabilities that convection causes in fluids. These guides allowed us to make predictions that could to some extent be checked.

The hot-spot theory

Toward the end of the 1960s, the nascent theory of plate tectonics at last provided a logical framework to explain the geographical distribution of most earthquakes and volcanoes, such as those of the mid-ocean ridges and subduction zones. Now we realized, the volcanoes around the Pacific's "Ring of Fire" were situated just above zones where one plate plunged under another, to submerge itself in the mantle. Submarine volcanoes, far less spectacular than those of the famous ring, punctuated the 60,000 km of mid-ocean ridges where new crust was being created and the plates spread apart; these volcanoes emerged only at exceptional points like Iceland and Afar. In short, volcanoes and earthquakes were concentrated at the boundaries between plates and even served to define them – but were not seen at the center of plates.

1 See for example V. Courtillot and G. Vink, How continents break up, *Scientific American*, 249, 42–49, 1983, or a more recent but more technical paper by I. Manighetti, P. Tapponnier, V. Courtillot, S. Gruszow and P.Y. Gillot, Propagation of rifting along the Arabia–Somalia plate boundary, *Journal of Geophysical Research*, 102, 2681–2710, 1997.

Yet Tuzo Wilson had noticed that some major volcanoes could not be explained by the new theory. The magnificent edifice of the Big Island of Hawaii for example, which is the largest volcano on Earth,[2] rising nearly 10,000 m or 33,000 feet above the surrounding ocean floor, is thousands of kilometers from any active plate boundary. Noting that the island was the tip of an archipelago that sank gradually into the ocean and extended off to the northwest, Wilson had the idea, guided by the absolute ages that the Australians had found for some of the islands, that an abnormally hot area must be located deep down in the mantle, burning its way through the overlying plate like a blowtorch to form a volcano. As the plate passed over this hot spot, a line of volcanoes formed, each older and thus cooler and denser than the next, sinking beneath the waves as the zone that remained active moved farther and farther away (Fig. 5.2).

This theory has been considerably expanded and generalized by Jason Morgan, undoubtedly one of the most important of the founders of plate tectonics. Morgan found a few dozen hot spots, analogous to Hawaii, and studied the relative positions of the chains of extinct volcanoes they had created. He showed that these hot spots, which he believes are anchored very deep in the mantle, have moved little relative to one another over time. Hence the four main hot spots that pierce the Pacific Plate – Hawaii, Easter Island, McDonald and Louisville – draw four enormous parallel chevrons there (Fig. 5.2): the archipelago of the Hawaiian Ridge plus the Imperial Seamounts, the Tuamotu Archipelago with the Line Islands, the Cook Islands and Austral Ridge with the Marshall and Gilbert Islands, and, finally, the Louisville Seamount Chain. These submarine volcanoes have been dredged and their specimens dated, primarily during the missions of the *Glomar Challenger* and later of the *JOIDES Resolution*:[3] vessels equipped for the international deep-sea drilling program. The results confirmed the theory magnificently: for example, for the trail left by Hawaii, the ages regularly grow older as you go farther to the northwest until they reach around 40 Ma at the great elbow point at

2 But not the largest in the solar system. The trophy for that goes to a volcano of the same type, Nix Olympica, also called Olympus Mons, on the planet Mars. It has a base diameter of 600 km and an altitude of 26,000 m, or about 85,000 feet.

3 Joint Oceanographic Institutions for Deep Earth Sampling, an international consortium of universities collaborating in marine research. *JOIDES Resolution* is their main vessel.

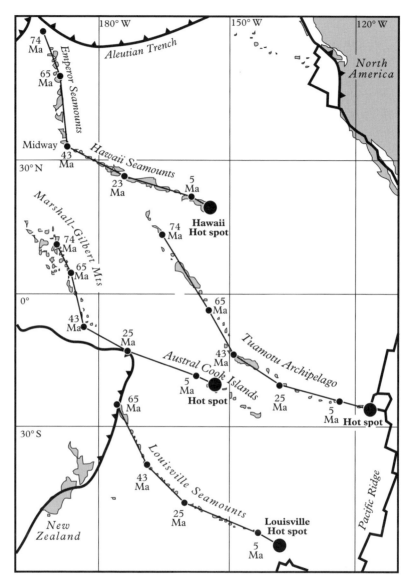

Figure 5.2
Pacific hot spots and their associated chains of seamounts, with ages indicated. The part of the
Hawaiian chain older than 75 Ma has been subducted in the Aleutian trench.

Midway. Then the archipelago turns away to the north. The last dated
volcano to the north is 70 Ma old. The oldest part has been swal-
lowed up under the Asian Plate, where the Kurile subduction trench
joins that of the Aleutians.

Subsequent to our work in the Deccan, a *JOIDES Resolution* mission was decided upon. Led by Robert Duncan of the University of Oregon at Corvallis, it dated the underwater mountains that link Réunion to India (Fig. 5.1). The volcanism of what was once called "Bourbon Island" has been active for 2 Ma. Before Piton de la Fournaise, it created Piton des Neiges farther to the north, now extinct and cut by erosion into three magnificent cirques. Before Réunion was born, the volcanism of the island of Mauritius began, 7 Ma ago. Farther north, the ancient volcanoes are now under water. The ages of the rocks sampled from the sea floor increase regularly from 35 Ma south of Saya de Malha to 48 million in the Chagos, reaching 55 to 60 million north of the Maldives, just before meeting up with the traps. The elegant regularity of the ages and morphology of the archipelagoes is interrupted only because the hot spot, originally located under the Indian Plate, crossed the Mid-Indian Ridge some 40 Ma ago and, therefore, passed under the African Plate. The older crust of the African Plate bears the most recent imprint of its "burn."

Plumes and instabilities: from honey to the mantle

Unlike the trail of Hawaii, whose origin is forever lost in the mantle, no subduction has obliterated the ancient history of the Réunion hot spot. The Deccan Traps are the evidence left by its birth. In this insight we are guided by some very fine experimental results. Morgan had already suggested that the hot spots corresponded to another of Earth's dynamic regimes, a different mode of convection from the more regular and broader-based system of plate tectonics. Material from the mantle, hotter and thus lighter than its environment, was thought to grow unstable and rise rapidly through the cooler remainder of the mantle, which was denser and more viscous.

Unfortunately we cannot look inside the Earth, except very fuzzily and indirectly by studying the gravitational field, the magnetic field, or, with just a bit more precision, the propagation of seismic waves. Seismology has recently made considerable advances, and in the late 1990s, thanks to a kind of tomography analogous to the CAT scans used in medicine to obtain images of the interior of the human body, we have begun to "see" the hotter or less hot zones of the mantle in

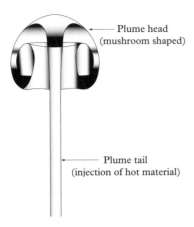

Plume head
(mushroom shaped)

Plume tail
(injection of hot material)

Figure 5.3
The head and tail (or
stem) of a "creeping
plume" (after Hill),
rising as an instability
in a fluid mechanics
experiment.

three dimensions. In this way we can distinguish abnormally hot
zones under Hawaii, Iceland, or Afar, but still with too little resolution and spatial precision to detect the thin plumes that would open
up into these hot spots.

So let us turn to analogy instead. In fact, we can attempt to repro-
duce in the laboratory, on a reduced scale, the evolution of a thin
layer of fluid lighter and less viscous than the fluid lying above it. The
first experiments of this type were performed in 1975 by Jim
Whitehead, of Woods Hole, and were later resumed with a very wide
variety of fluids ranging from water to honey by such researchers as
Peter Olson in Baltimore, Ross Griffiths and Ian Campbell in
Canberra, and David Loper in Florida. The latter placed a thin film
of silk on top of a container full of a viscous syrup, and on top of
that placed a layer of water tinted with ink. Then he inverted the container. The water, less dense and less viscous, was then in an unstable position. The silk film acted as a retardant and filter and prevented
the water from rising instantaneously to the new top. After some time,
pockets of water began to grow and rise in the syrup, somewhat in
the manner of what geologists call diapirs (Fig. 5.3). This is how salt
diapirs, light and plastic, rise through layers of rock closer to the surface, sometimes facilitating the formation there of traps where petroleum and gas gather. As our little diapirs of colored water rise through
the denser fluid, their heads assume a shape like the caps of some
kinds of young mushrooms. The upper part of this head is nearly
spherical. It is the site of an internal convection that is responsible

for the rolled-up appearance of the lower part of the head. The head of the plume remains connected to the lower layer by a thin umbilical cord, through which it is fed and grows. Fluid mechanics has dubbed these objects "creeping plumes."

Head and tail

Pursuing an idea first put forward by Jason Morgan in 1981, in 1986 Jean Besse and I proposed that the Deccan Traps should be viewed as the first appearance of the head of the hot spot now located beneath Réunion[4] (Figs. 5.1 and 5.3). The size of this head[5] would explain the magnitude of the dynamic effects and volume of the volcanic eruption, in excess of a million cubic kilometers. Basalt results from the melting of only a small percentage (less than ten) of the rocks in the mantle. So at its birth, the head must have had a diameter of more than 400 km , or nearly the full thickness of the upper mantle. This head constituted a formidable heat source. It caused the Indian Plate above it to arch in a vault and grow thinner. Making its way through a cold, ancient crust of granitic (acidic) composition, rich in volatile elements, and finally traversing sediments, it at first gave birth to an explosive volcanism. Then, after complete fracturation had set in, the great basaltic floods were successively laid down.

In the case of the birth of the Iceland hot spot, hundreds of layers of basaltic ash have been found as much as 2000 km away from the emission points, indicating the force of the first explosive eruptions. The modifications of the upper mantle and the Earth's crust in the case of the Deccan Traps were such that the mid-ocean ridge then operating in the Indian Ocean, far to the south of India, stopped widening and "jumped" farther north, tearing away from India the piece of continent that today emerges only in the Seychelles (Fig. 5.1). Thus the emergence of the head of the hot spot led to the opening of a new ocean, the basin that became the Arabian Sea. But the eruption of the traps emptied the head of the plume somehow, and the Indian Plate, drifting farther north, now moved over the plume's

4 In 1984, José Achache and Philippe Patriat, of our team, had reconstructed the trail of the hot spot from the kinematics of the Indian Ocean and found that the point had been located near the Deccan traps around the time of the KT boundary.

5 By theoretically expanding the laboratory experiment to the scale of the Earth's mantle.

tail (or umbilical cord), thus generating the far less expansive archipelagoes that lead toward the plume's modern opening at the surface, Piton de la Fournaise.

The Deccan is not alone

This model of a plume with a dual structure – a voluminous head and far thinner tail – has been generalized by Mark Richards, Bob Duncan, and myself to cover all active hot spots on the globe, or at least those that have not lost their heads under a subduction zone, as Hawaii may have (Fig. 5.4).[6] So the hot spot now located under Yellowstone in the USA was born some 16 Ma ago in the form of the great volcanic province of the Columbia River, located only a few hundred kilometers to the west and 30 Ma ago, the Ethiopian Traps marked the arrival at the surface of the hot spot that still remains not far away, in the Afar region, because the African Plate has moved little relative to the mantle. Even longer ago, 57 Ma, the hot spot of Iceland built the enormous volcanic stacks that today form the cliffs of the east coast of Greenland and the entire northwest margin of both the British Isles and the Norwegian continental plateau. In this case the traps were riven shortly afterwards (like the Seychelles and the Deccan) by the propagation of the Mid-Atlantic Ridge, which would give birth to the North Atlantic Ocean. The lesser ridges formed by the hot spot are short and join King Christian IX Land and the Færoe Islands to Iceland. The exceptional emergence of a mid-ocean ridge in Iceland, and also in the Afar region, occurs because here a rather stable hot spot joins forces with normal sea-floor spreading, thus intensifying both lava production and thermal effects.

Well before the Deccan Traps were formed, around 135 Ma ago, the great volcanic expanses of Parana formed in South America, where erosion has since created the superb Iguaçu Falls. The Parana Traps can be linked to the hot spot that emerges at the island of Tristan da Cunha in the South Atlantic. Situated right on the mid-ocean ridge, this point also created the Walvis Ridge to the east, on the African Plate. And this latter ridge in turn leads to a small outcropping of

6 See also the work of the Australian National University group, for instance I.H. Campbell and R.W. Griffiths, Implications of mantle plume structure for the origin of flood basalts, *Earth and Planetary Science Letters*, Vol. 99, 79–93, 1990.

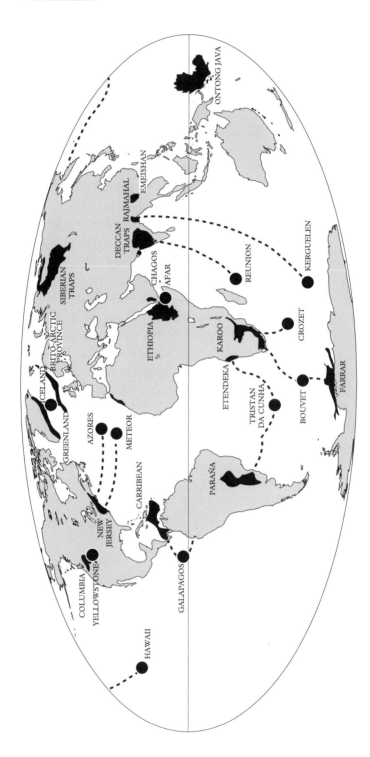

Figure 5.4
World map of the main traps (or flood basalts). Some have been linked to the currently active hot spot volcanoes, whose birth may be the cause of the traps.

traps on the coast of Namibia (Etendeka), which was separated from the main body of the Parana basalts by the opening of the South Atlantic shortly after both parts were laid down. We can also cite the gigantic submarine Ontong–Java Plateau in the Western Pacific, born perhaps 110 Ma ago with the Louisville hot spot. The Karoo lavas in southern Africa and the Farrar lavas in Antarctica, 184 Ma old, can be linked to the Bouvet hot spot. Other events are thought to have created the Rajmahal Traps in India and the traps along the western margin of the North Atlantic and the Jamaica Plateau. Finally, the immense Siberian Traps may be associated with the hot spot of Jan Mayen, a little north of Iceland. We will come back to these last, because they are of unusual importance in our story.

The birth of hot spots and continental breakup

So about ten volcanic episodes of exceptional magnitude have occurred over the past 300 Ma of the Earth's history. Today the volume of their outpourings still measures in the millions of cubic kilometers; in the case of Ontong–Java, for example, it may reach some ten million cubic kilometers. Pulled together in the simple shape of a sphere, the head of abnormally hot material rising from the mantle, whose partial melting created the lava, may thus have reached a diameter of over 700 km, which might have brought the lower part of the mantle into play. In most cases, the propagation of a break in a continent and the appearance of a new ocean basin seem to have been the result: the recent opening of the Gulf of Aden and the Red Sea (Afar), the North Atlantic (Iceland), the Arabian Sea (Deccan), the South Atlantic (Parana), the Southwest Indian Ocean (Karoo) and the Central Atlantic (New Jersey).

There often seems to be a direct relation between the birth of a hot spot and the breakup of a continent. When the continental crust thins out and breaks apart to give birth to an ocean above a mantle at "normal" temperature (around 1300°C), a typical oceanic crust forms, basaltic and some 7 km thick. The rocks of the upper mantle are rapidly decompressed, without losing their heat. So they begin to melt.[7] This decompression may be quite rapid on the geological time

---

7 This is called melting by adiabatic decompression.

scale, taking only about a million years. Dan McKenzie and Robert White, of Cambridge, have proposed a quantitative formulation for it: the longer the decompression phase takes, the more the upper mantle will cool and the less abundant the volcanism will be (some of the molten rock has a chance to solidify in the depths, before rising). In extreme cases, the oceanic crust may be no more than 2 km thick.

If the break occurs above an abnormally hot mantle, for example because of the presence of a hot spot (whether nascent or already established), the quantity of molten material may quadruple.[8] Only one-quarter of this liquid reaches the surface, where it is found in the form of a surface layer of basalt several kilometers thick. The remaining three-quarters remains underplated in the depths under the crust or is injected at the base of the crust, where it slowly cools.

The anomaly in heat and density represented by the head of a nascent plume may have the effect of lifting the continental crust, which bulges and forms a vault or arch that may exceed 2000 m in height, across more than 1000 km. This uplift may cause the entire region to emerge from the water. The combined effect of the potential energy thus supplied, and possibly also secondary convection within the head of the plume as it "digests" the lower part of the continental crust, places the crust under stress, and it grows thinner. This thinning further increases the melting caused by decompression of the mantle and may help to create the fissures through which the magma can escape. Breakup becomes more likely if the plate is already under tensile stress because of boundary conditions at its remote edges, such as downward pull at subduction zones. The resulting flows may then descend the slopes over long distances, as in the Deccan Traps: a slope of one per thousand would have been enough for them to spread out over hundreds of kilometers within a few days or weeks. In some cases, the preexisting topography does not allow flows to spread far. This was undoubtedly the case at the birth of the Iceland hot spot: many small sunken basins that had formed between today's northwestern Europe and Greenland limited the expansion of the flows, much of which then spread along the continental margin, including under the sea.

8 This hot spot is responsible for a temperature increase (really quite modest) of 200°C, or in relative terms, a thermal anomaly of only 10 to 15%.

A small debate went on for some time between Mark Richards, Bob Duncan, and myself on the one side, and Dan McKenzie and Bob White on the other. The latter, authors of an elegant quantitative model that makes it possible to calculate the quantity of emitted basalt as a function of the temperature of the mantle and the thinning of the lithosphere, leaned toward the idea that the breakup of continents and the emission of traps are two independent phenomena, which may occur together but merely by coincidence. We, on the other hand, thought them linked by a strong causal relationship: the birth of a hot spot under a continent generally leading to fissuring followed by the birth of a new ocean basin. The resolution of the debate lay in the precise measurement of the ages of the first flows, the first signs of decompression, and the first ocean floors. For the same reasons that we encountered in dating the Deccan lavas, it is very difficult to take this measurement with the necessary precision, and we are still far from having sufficient data in all cases.

Some plumes do not result in continental breakup, as in the case of the Columbia River, Hoggar in North Africa, or, possibly, the Siberian Traps. But in most cases when a new ocean formed, the major part of the eruption of the traps did indeed precede or coincide with a spreading action and the formation of the first sea floors. This was the case with the Deccan, Parana, Greenland, Ethiopian, Karoo, and other traps. I would still say that the first and shared cause of both traps and continental breakup is the arrival, under the lithosphere, of the plume from the mantle. But although there would be no breakup without the emergence of the plume head, it is also necessary for the lithosphere to be in a proper state of stress owing to remote plate boundary forces.

This view sheds a rather different light on the organization of continental drift. Traps mark the sites where many major ocean basins opened and, therefore, govern the future shapes of large oceans (at least those oceans that are not bounded by destructive subduction zones). For instance, the geography of the Atlantic Ocean, which comprises three large basins (North, Central, and South), reflects the impact points of three mantle plumes (Iceland, Central USA, and Tristan da Cunha).[9] Under this interpretation, plumes result from

9 See V. Courtillot et al., On causal links between flood basalts and continental breakup, Earth and Planetary Science Letters, 166, 177, 1999.

an anomalous, unstable mode of mantle convection, which leaves an indelible imprint on continents and paves the way for sea-floor spreading – the more normal mode of mantle convection – which then ensues. Some time ago, seismologist Don Anderson of Caltech proposed that long-lasting supercontinents would prevent heat from escaping from the mantle to the surface, thus promoting heat accumulations. Claude Jaupart and Laurent Guillou, of the Institut de physique du globe de Paris, have conducted pertinent experiments showing how plumes can form under a supercontinent and be drawn by convection to the central zone underlying the continental masses. This would readily explain the pre-breakup position of traps.

The extent of the traps on the surface may be greater than the size of the head of the hot spot within the mantle. However, the millions of cubic kilometers of lava emitted in the greatest traps are equivalent, as we have seen above, to a volume of mantle material that may have attained around 100 million km^3, spanning the entire thickness of the upper mantle.

An episodic, chaotic mechanism?

Like many other matters, the depth of the origin of hot spots is an object of debate (Fig. 5.5). Here I defend the idea of a deep initial origin (see Chapter 7). But others prefer an intermediary origin or even, like Don Anderson (who admittedly belongs to a small minority), argue for a shallow origin just under the lithosphere. But at any rate, plumes are fascinating geodynamic objects. I believe they are born from the instability of a deep boundary layer – 670 km down at least, and perhaps as deep as 2900 km (see Fig. 5.5) – and are essentially random. And so they are immediately reminiscent of the so-called intermittent regime suggested by writers on chaos theory.[10] The appearance of plumes at the surface in the form of traps would, therefore, be an unpredictable event.

If the Deccan Traps had such consequences for life on Earth, what would be the situation for the other traps, so few in all, that we have just seen on our geological tour? Did they also steer evolution catastrophically in new directions?

10 For example, see P. Berger, Y Pomeau, and C. Vidal, *L'Ordre dans le chaos*, Paris, Hermann, 1984; and James Gleick, *Chaos: making a new science*, New York, Viking Penguin, 1987.

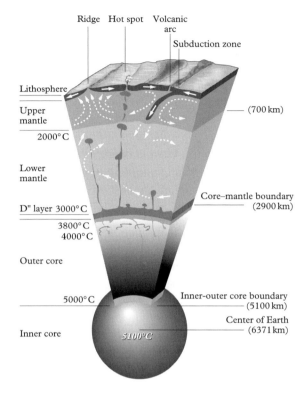

Ridge Hot spot Volcanic
                    arc

Subduction zone

Lithosphere

Upper   —— (700 km)
mantle

2000°C

Lower
mantle

Core–mantle boundary
D" layer 3000°C   —— (2900 km)

3800°C
4000°C

Outer core

5000°C   Inner-outer core boundary
                    —— (5100 km)

                    Center of Earth
Inner core   —— (6371 km)
            5100°C

Figure 5.5
A schematic cross-
section to the center
of the Earth, incor-
porating some of
the ideas described
in the text, particu-
larly mantle (and
core) plumes,
inspired by fluid
mechanics and
numerical
experiments.

6

A remarkable correlation

We have devoted a considerable portion of the earlier chapters to the disappearance of the dinosaurs and ammonites in the KT crisis. Yet this is not the most impressive mass extinction in the past 600 Ma. The greatest one occurred at the end of the Paleozoic Era, 250 million years ago (see Chapter 1). This was the end of the trilobites. Most marine life forms were severely affected; fish and shellfish were decimated. Nor was the situation much better on land. Proto-mammals, amphibians, and reptiles disappeared en masse, and no doubt it was a close call as well for one of the few surviving species of proto-mammals, the one from which we would eventually descend. Our presence on Earth hangs only by this slender, very ancient thread. In all, nearly 95 percent of species disappeared in about 2 Ma, or even less if we can believe the recent work of the Chinese scientist X. Yang and the American S. M. Stanley.

The mother of all extinctions

At the end of the Paleozoic, almost all the continents had finished gathering together to form Pangea.[1] An elongated ocean opened to the east in this "supercontinent": Tethys Sea, the ancestor of the Mediterranean. Its floor has since largely disappeared in the collisions between Africa, Arabia, and India, on one side, and Europe and Asia, on the other, so that now remnants are found in the Alpine chains that extend from the European Alps to the Himalayas. In the

1 Pangea was the single continent, gathering together most of today's continents, that seems to have existed as a distinct entity between about 300 and about 200 Ma ago. Wegener introduced the concept and the name, although he did not know the more ancient phases that led to the agglomeration of the supercontinent. About 200 Ma ago, Pangea began to break up into Gondwanaland to the south and Laurasia to the north as the Central Atlantic Ocean opened. Then, the nascent Southwest Indian Ocean split up Gondwanaland itself into two parts: eastern Gondwanaland (consisting of India, Australia, and Antarctica) and western Gondwanaland (Africa and South America).

midst of Tethys Sea new sea floor was forming, while at its margins, particularly the northern ones, a succession of subductions and collisions prevailed. Unfortunately, relics of this period are rare and often poorly preserved. All the ocean floors that existed at the time have disappeared. Some have returned to the mantle, in the subduction zones; a few have been deformed and incorporated (as ophiolites) into mountain chains, which have then undergone erosion and sometimes new tectonic phases. Where sediments of the Permo–Triassic boundary are observable, the series is often incomplete. As it did at the end of the Cretaceous, the sea withdrew, interrupting deposition and leaving formations vulnerable to erosion. This marine regression seems to have been even more substantial than at the end of the Mesozoic: the sea level dropped 250 m. The phenomenon may have been linked to changes in the relative velocities of the crustal plates and to the considerable shrinkage of the shallow seas that existed when Pangea lay agglomerated in a single, gigantic continental mass.[2] Withdrawal of the seas causes profound modifications in environments and is often enough in itself to put species sorely to the test.

Some of the best outcrops from the Permo–Triassic boundary can be found in China. Here the mass extinctions seem less selective than at the KT boundary. Organisms living on the sea bottoms, at the continental margins, and in shallow waters were massively wiped out:[3] fusulinids (giant Foraminifera characteristic of Permian sediments throughout the world), reef-building organisms, crinoids ("sea lilies"), bryozoans and brachiopods, and almost all the nautiloids disappeared. The ammonites, which would dominate the Mesozoic seas, owe their existence only to the survival of a few rare species of these cephalopods, the goniatites. The rare sections that reflect the continental environment, preserved in China and South Africa, show chaos among the plant species and carnage among mammal-like reptiles: only about one genus in a hundred survived.

Very recently, Stanley and Yang have focused on taking a minute inventory of genera from six marine groups existing throughout the Permian Period: brachiopods, ammonoids, bryozoans, fusulinids,

2 Recent work by Jean Besse and Frédéric Torcq   3 See Chapter 1, Note 10.
has shown that Pangea probably suffered major
deformation along a huge shear zone between
250 and 200 Ma ago.

gastropods, and bivalves. They found two brief, intense crises, one at the end of the Guadalupian (the next-to-last-stage of the Permian), and the other 8 Ma later, at the end of the Tatarian (the last stage of the Permian and, therefore, also of the Paleozoic Era). Not only was the Permo–Triassic crisis the greatest of all time, but, uniquely, it had been preceded not long before by another crisis that was almost as intense. This explains why, to the first observers, it seemed spread out over time. Life on Earth no longer looks the same after this double crisis. It would take the Earth's ecosystems several million years to recover.

The analysis of the physical, chemical, and isotopic variations in the sediments from the few available sections of the Permo–Triassic boundary was summarized by the late Israeli geochemist Mordekai Magaritz and his colleagues. The clearest information comes from the distribution of carbon isotopes. It seems that a rapid and major drop in the carbon isotope ratio[4] may indicate that organic (dead) carbon was deposited in abundant quantities, stored, and then oxidized, possibly because of the erosion that resulted from the regression of the seas. This would reveal the traces of a decline in biological productivity, together with a drop in the atmosphere's oxygen content and an increase in its carbon dioxide. The elements of the platinum series, particularly iridium, offer no indication of the presence of extraterrestrial matter, nor does the composition of clays or the rare-earth content of the formations. There is not the slightest amount of shocked quartz, nor apparently a single extraterrestrial microspherule. The discovery of elevated iridium concentrations (from 2 to 8 p.p.b.) by two Chinese teams in sections from Zhejiang and Sichuan provinces, announced around 1985, has remained unconfirmed by the many careful measurements taken by three different groups, led, respectively, by Frank Asaro, Carl Orth, and Robert Rocchia, who observed only values some 50 to 1000 times lower! Apparently the Chinese teams' samples were contaminated.

4 Carbon has two stable isotopes, ^{12}C and ^{13}C, which are fractionated by inorganic and organic reactions. In general, carbonates are enriched in ^{13}C while organic matter is enriched in ^{12}C. The difference between the $^{13}C/^{12}C$ ratio of a speci- men and that of a standard, expressed in parts per thousand, is called $\delta^{13}C$. This ratio gives an idea of the magnitude of the biomass present on Earth, as recorded in a sediment of a given age.

The age of the Siberian Traps

In 1982, Jason Morgan had already suggested that this major extinction of the Phanerozoic might be associated with the massive volcanism that once covered part of Siberia. Our work in the Deccan Plateau led us, of course, to ask this simple question: were other traps associated with other major extinctions? To have any value whatever, a scientific model has to permit predictions. Its success is a function of its ability to withstand such tests. We had to start with the greatest extinction of all time; so we predicted that the end of the Permian must coincide with the Siberian Traps. Located at the northwestern margin of the Siberian platform, these traps today cover an area of 350,000 km^2. Their cumulative thickness reaches 3700 m in places, and the volume of these dozens of flows must originally have been in excess of 2 million km^3, just as in the case of the Deccan (and it may even have been much more, in view of their age and the erosion to which they have been exposed). The proposition we formulated, following in Morgan's footsteps, was easy to verify. Work similar to that done in the Deccan Traps should provide an answer.

The first geochronological work by Soviet researchers had suggested that this volcanism, covering continental sediments of the Upper Permian, had been spread out over 40 Ma. But in 1991, again using the argon-isotope method, Paul Renne at Berkeley and his colleague Asish Basu showed that the major portion of the traps had been laid down in less (possibly much less) than a million years, 248 Ma ago (with an uncertainty of 2 Ma for the absolute age). These dates have been compiled by several laboratories (including Corine Hofmann at our own laboratory), and using generally accepted standards cluster around 250 Ma. A compilation of magnetic data by Russian, American, and French researchers showed one main reversal of the magnetic field in the volcanic stack, again pointing to the fact that the event must have been rather brief.

Remarkably, Renne and his colleagues have been able to date two volcanic tuff layers from stratigraphic sections in China that straddle the Permo–Triassic boundary. Because they use the same technique and standards, they were able to show that the ages of the traps and of the boundary in China were virtually identical (with an uncertainty of only a few hundred thousand years).

Whereas the evidence was less clear for the Deccan Traps, the lava and ash that outcrop at the base of the Siberian Traps show that the eruptions were preceded or accompanied by highly explosive phases and by a heavy emission of sulfur, implying major climatic effects. The chemistry of these traps indicates a source with an original composition belonging to the mantle and contaminated by its passage through the ancient continental lithosphere of Siberia. A mantle plume was once again the designated culprit. It should be noted that the eruption of the traps involved neither the dislocation of the Asian continent nor the birth of a new ocean – though in fact, the traps are bounded to the west by the Kazakhstan Basin, filled with more than ten kilometers of sediments. Below these sediments there may well lie the traces of an aborted ocean, a failed attempt of the Asian crust to break up.

It's worth recalling that Stanley and Yang provided tantalizing evidence that the late Permian extinctions might actually have been a "double whammy," involving an earlier event some 8 Ma before the Permo–Triassic boundary proper. And indeed, there is yet another occurrence of flood basalts, located in southern China, little known and poorly preserved: the Emeishan Traps. These have been extensively damaged by the southeastward extrusion of Indochina, a secondary consequence of the collision of India into Eurasia. Recent paleomagnetic results by Neil Opdyke and Kainian Huang of the University of Florida show that the traps erupted fast. And stratigraphy demonstrates that this happened just at the end of the Guadalupian. So the Emeishan Traps are an excellent candidate for the first of the two late Permian mass extinctions.

The duration of the Permo–Triassic events would have to be considered in relation to that of the two volcanic eruptions and to the duration and intensity of the (possibly related) shrinkage of the seas. We have seen that the (double?) drop in sea level was accompanied by a drastic shrinkage of the shallow seas that bordered the continents and provided shelter for extremely diverse communities. There seem to be a great many similarities between the events that accompanied the two greatest crises of the past 300 Ma: rapid eruption of enormous continental basalt formations coincides with two mass extinctions, but so far without any clear indication of an impact at the end of the Paleozoic.

The end of the Triassic, volcanism, and the opening of the Central Atlantic

In any event, the "volcanic" model had won an important victory. But what about the other traps and other extinctions? The crisis that separates the end of the Triassic from the first stage of the Jurassic, the Hettangian, is one of the five great crises of the Phanerozoic (see Fig. 1.2, p. 11). This crisis, 200 Ma ago is generally considered somewhat less intense than the KT crisis, although some types of measurement of extinctions suggest the two are almost equal. During this time, the continents were still gathered together in the great mass of Pangea, and the end of the Triassic saw a phase of marine regression. The low sea level once again explains the limited development of seas along the continental margins, as well as the very small number of detailed and continuous geological sections that have preserved the record of the event.

It appears that towards the end of the Triassic almost all genera of ammonites became extinct. They were accompanied into oblivion by more than half the genera of bivalves, representing almost all of their species, and many brachiopods and gastropods (more even than at the Permo–Triassic boundary). The conodonts,[5] which had reigned through the Paleozoic and managed to survive the Permo–Triassic boundary, finally became extinct. The corals and sponges, and the entire reef ecosystem they constitute, collapsed and did not reoccupy the world ocean, in new forms, until 10 Ma later. The Foraminifera, less thoroughly studied than those of the Cretaceous–Tertiary boundary, seem to have been less severely affected: "merely" 20 percent of the families disappeared. On land, 80 percent of the quadrupeds became extinct, and the flora underwent a major reorganization. Whereas below this boundary the dinosaurs were still relatively few and fairly small in size (up to "only" 6 m long), they diversified very rapidly afterward: some species grew more imposing in size and represented as much as 60 percent of the population of which we have fossils. One might say that the Triassic–Jurassic boundary marked the real beginning of the reign of these species.

---

5 These fossils are one of the primary tools of fine stratigraphic correlation for the Paleozoic and the Triassic. Their color under the micro- scope depends on the temperature to which they were exposed and helps to determine the conditions under which petroleum matures.

The duration of the crisis is poorly understood, for lack of sufficiently precise and numerous – and sufficiently well-studied – records. The other boundaries have not enjoyed the amount of energy that researchers from all disciplines had lavished on the KT boundary. British paleontologist Anthony Hallam cites figures of less than a million years. The Manicouagan Crater in Quebec, 70 km in diameter, was created by the impact of a meteorite (or comet); and for a time, this crater was thought to be a possible origin of the late Triassic extinctions. But a precise dating to 220 Ma ago, nearly 20 Ma before the boundary, now seems to rule out any cause–effect relationship. Finally, following a recent spate of more intense research, traces of shocked quartz are claimed to have been found in an Italian section. However, few propose an extraterrestrial origin for the Triassic–Jurassic boundary.

However, very abundant volcanism did occur around this period. Basalts located in West Africa, and especially the abundant series in the eastern North American continent,[6] coincide rather precisely with the boundary, within a precision better than 50,000 years, according to Paul Olsen from Lamont. These include the Palisades Sills, known to anyone who has crossed the George Washington Bridge on the way out of Manhattan. A pollen study has shown that the paleontologic boundary and the volcanic episode virtually coincide. Moreover, this episode corresponds with the first great phases of the breakup of Pangea, heralding the opening of the Central Atlantic Ocean. The stable isotopes, such as those of oxygen, do not seem to have recorded any major change in climate. However, the drastic changes in sea level in a world without ice caps, and traces of anoxic events in the ocean, are compatible with an internal origin of the biological crisis. Hallam sees these as the consequence of abnormal mantle-plume activity and the resulting intense volcanism.

Other traps, other extinctions

Over the past five years, other traps have been studied in detail. In almost every case, these objects, measuring millions of cubic kilo-

6 These two regions were contiguous at the time, since the Central Atlantic had not yet opened up.

meters in volume, were laid down within a very short geological interval, on the order of a million years or sometimes even less. This is
the case of the North Atlantic Tertiary Volcanic Province, which
marks the birth of the Iceland hot spot and the opening of the North
Atlantic. The Greenland Traps, over 3000 m thick in places, were
laid down in two pulses centered at 59 and 56 Ma. Both pulses seem
to have recorded only one (different) reversed period of the magnetic
field. The second pulse coincides in age with the boundary between
the Paleocene and the Eocene, where the extinctions admittedly are
minor.

The age of the enormous Parana Traps in Brazil (Fig. 5.4, p. 82)
has just been gauged by Paul Renne at 133 Ma, with an uncertainty
of 1 Ma.[7] Preliminary paleomagnetic results once again seem to show
only one magnetic-field reversal there, bearing witness to the event's
brevity. The Jurassic–Cretaceous[8] boundary is not known precisely
(values between 130 and 145 Ma have been proposed). Although the
older ages are often preferred, the younger ages are compatible with
the age of the Parana Traps. These traps (also called the Serra Geral
Traps) mark the birth of the Tristan da Cunha hot spot and were
followed by the opening of the South Atlantic. Moreover, a part of
the traps, Etendeka, remained attached to the African continent.

The Madagascar volcanism, 90 Ma ago, may mark the end of the
Cenomanian, while that of Rajmahal, in India, 116 Ma old, could
mark the end of the Aptian and the birth of the Kerguelen hot spot.
The Farrar lavas in the Antarctic, whose emission coincides with the
main eruptive phase of the Karoo in South Africa, about 180 Ma ago,
do not clearly match up with any major extinction. They began to
erupt when the ocean separating the two major parts of Gondwanaland began to open up.

Recent work has revealed uplift of the continental basement, rift-
ing and extensive volcanism in the Kola, Vyatka and Pripyat–
Dniepr–Donets provinces, spanning almost 2000 km on the East
European platform in the former Soviet Union. Drilling and seismic

7 This conclusion, however, is the subject of an
ongoing controversy with a largely British team
of researchers who think the volcanism may have
spanned 10 Ma.
8 Christian Koeberl and colleagues (*Geology*, 25,
731, 1997) have discovered and drilled a potential impact site in South Africa. They find evidence for shock and extraterrestrial material at
an age of 145 million years, close to the more
generally accepted age of the Jurassic–Cretaceous
boundary.

data from the Pripyat–Dniepr–Donets area demonstrate that this rifting and intense volcanic activity took place in the Late Devonian.
Two distinct events are recognized, at the end of the Frasnian (near
365 Ma) and the Famennian (near 355 Ma), respectively. This is the
period of two of the most important extinction events in the Paleozoic
(see Figure 1.2, p. 11), the oldest occurrence of a connection between
trap volcanism, continental rifting and mass extinction that I am
aware of.

Under a joint Franco-Ethiopian cooperation program, our group
has completed a study of what may be the least known yet most recent
traps. Before the Red Sea and Gulf of Aden opened up, the large
Ethiopian Traps formed a single mass together with those located in
South Yemen. Corine Hofmann, Gilbert Féraud, Pierre Rochette
and I, working with Ethiopian colleagues led by Gezahegn Yirgu,
have just finished a magnetostratigraphic and geochronologic analysis of a beautiful section of basaltic lava, 1700 m thick, near Lima-
Limo on the northern edge of the plateau. We found only two
magnetic reversals, and the age is very close to 29.5 Ma. So like all
the others, the Ethiopian–Yemen traps erupted over less than 1 Ma.
However, this is not the time of the Eocene–Oligocene boundary, as
I and others had suggested before. Interestingly enough this boundary, at 33.7 Ma, was something of a nonevent, as paleontologists such
as William Prothero tell us. Some extinctions did occur around
37 Ma, and Alessandro Montanari and his colleagues have found an
iridium anomaly in Italian sections at this time. But the main clima-
tological events occurred very near 30 Ma ago, at what has been
defined as the Lower to Upper Oligocene transition. This was the
time of the largest drop in sea level – on the order of 100 m – between
the beginning of the Cenozoic and the declines related to recent
glaciations. It was also a time of cold, extremely dry weather, and a
time when large-scale continental glaciation began in the Antarctic.
Finally, it was a time of extinctions and minimal biological diversity.
So our new results reveal yet another instance in which a brief,
catastrophic eruption of traps coincides with a major event in the biosphere.

What's more, this is also the moment when Arabia and Somalia
split away from Africa along the Red Sea and the East African Rift.
A short time later, rifting began along the Gulf of Aden, which is

now propagating a tear in present day Afar and severing Arabia from
Africa as the East African Rift dies. It seems particularly significant
that the rifting propagated toward the zone of the former plume head,
not away from it. Isabelle Manighetti, Paul Tapponnier, and I are
still studying the phenomenon.

Because the African plate moves so slowly over the mantle hot
spots, and because this is the youngest major plume event, the plume
tail (Afar) is still close to where the plume head erupted (the
Ethiopian traps): the three rifts that meet in Afar have not yet spread
wide. All of which makes this a particularly complex but attractive
area, where all these processes interact and some of the ideas I have
put forward here can be better tested in the near future.

Correlation

The most recent determinations of the ages of the 12 principal
basaltic provinces and the ten major extinctions that have occurred
over the past 300 Ma have been compiled by several authors.[9] Fig. 6.1
shows these findings: the correlation is almost perfect. Different
authors would no doubt propose slightly different lists and ages, but
it is unlikely that their conclusions would diverge much. The greater
the precision of the age determinations, the better the correlation has
proved to be. There is less than one chance in a hundred that such
a sequence could be random. As Jason Morgan suggested, once again
leading the way, followed more recently by Michael Rampino,
Richard Stothers, and myself, the ages of the traps coincide with the
principal divisions of the geological time scale, which themselves are
founded, as we saw in Chapter 1, on the principal phases of species'
extinctions. There are few exceptions, and the precision of these
determinations is quite high, ranging from 1 to 5 Ma, depending on
the case. So the dusty old scale we inherited from the nineteenth century actually does represent one of the great internal rhythms of the
Earth.

Two basaltic provinces (Columbia and Karoo) seem not to be
associated with an extinction. And two extinctions (those of the

9 See for example V. Courtillot, Mass extinc-   of the Earth Sciences, Jerusalem, 43, 255–266,
tions: seven traps and one impact? Israeli Journal   1994.

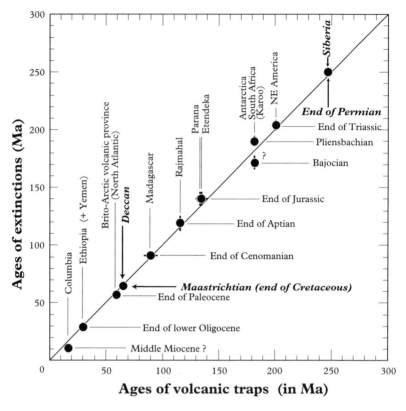

Figure 6.1
A comparison of the ages of the main traps (measured in most cases with potassium–argon or argon–argon geochronology) and mass extinctions (keyed to the geological time scale). Uncertainties shown as bars (or not shown when smaller than dot). All but two show an excellent correlation. The main Periods and Epochs are shown.

Pliocene and Middle Miocene) do not correspond with a trap. We may note, however, that the basalts of the Columbia Plateau, which are by far the smallest province of this type, might be related to some events in the Middle Miocene. Among the 12 traps younger than 300 Ma, at least nine can be associated with a major extinction. Seven of the ten principal extinctions can be associated with an episode of massive basaltic volcanism. As for the Pliocene extinction, many authors consider it relatively minor and of only regional importance; it may be linked with glaciations, or even the presence of the human species (see Chapter 10). At any rate, there is a striking association among the birth of a hot spot, the formation of basaltic traps, mass extinctions, and, generally, continental breakup and the birth of an ocean. By contrast, let us recall that even most partisans of the asteroid impact hypothesis, and particularly the iridium specialists, would agree that the only well-documented case in its favor is the KT boundary.

If we assume the climatic scenario associated with the volcanic hypothesis, clearly the magnitude of the biological effects will depend on a large number of factors: the arrangement of the continents, the sea level and climate at the time of the eruptions, the total amplitude, duration and number of the individual events, the closeness of these events in time, and so on. Javoy and Michard have shown the cumulative effect of some major flows that followed one another a few centuries apart. If we estimate the actual duration during which the major portion of the lava was laid down at only 100,000 years (or less), the mean flow of lava would be more than 10 km^3 per year, with convulsive phases that might be considerably more violent, such as the laying down of a flow of more than 1000 km^3 in less than a few weeks.[10] Even the lowest mean rates indicate that the laying down of traps is a geodynamic event of the utmost importance.

We have already emphasized the important role of the sulfur in the aerosols ejected into the atmosphere. Another important climatic parameter is the environment in which the volcanism takes place. This is no doubt the main reason why the enormous quantities of lava in the Ontong–Java Plateau, which erupted entirely under water over the course of about 3 Ma at the beginning of the Aptian, a little more than 115 Ma ago, had little biological impact; whereas the Deccan and Siberian Traps, belching forth through fissures into the open air as their gases were injected directly into the atmosphere, had devastating effects. In intermediate situations, basalts might erupt partly under water and partly in air: this is probably what happened with the traps of the North Atlantic Tertiary Volcanic Province, associated with the limited extinction at the end of the Paleocene.

A rapid glance at the time sequence of the traps might suggest a certain regularity. Rampino and Stothers, who in 1988 were the first

---

10 Compare this with the 12 km^3 in a year from the Laki fissure in 1783!

to offer a quantitative proposal that a correlation might exist between the ages of the traps and the extinctions, likewise noted that these events largely seemed to succeed one another periodically, about every 30 Ma. Four years earlier, the paleontologists Raup and Sepkoski believed they had shown a periodicity of 26 Ma between the major phases of extinctions of species. Should we view this as an indication of some fundamental clock? And should we look for the clockmaker in the sky, or underground? This question of periodicity has intrigued the Earth sciences' community for several years. We'll pause now to look at it for a moment.

7

Nemesis or Shiva?

Finding a periodicity in a series of observations taken over time calls for special techniques. Here we are getting into the field of "signal processing," which has seen extraordinary developments in the second half of the twentieth century, with the explosive proliferation of means of producing, transmitting, and storing information. But in order to extract the information concealed in a series of measurements spaced apart over time, particularly the more or less hidden periodicities, we still draw on methods whose foundations were laid early in the nineteenth century with the work of Joseph Fourier and Baron Prony.[1] Signal processing was conceived for analyzing signals that have a long duration in comparison with the periodicities being sought.[2] But the series formed by the occurrence times of the traps or extinctions are relatively short;[3] the data are generally anything but extensive, and the intensity of an event as well as its age are furthermore liable to serious uncertainty. So in this case, signal processing has led to some dubious reasoning and, on occasion, some overly self-confident statements of results.

A crisis every 30 million years?

Raup and Sepkoski had scarcely published the famous periodicity of 26 Ma when it came in for sharp attack. Several statisticians pointed out that the geological time scale itself – in other words, the succession of boundaries of geological periods in time – already contained

---

1 Several methods make it possible to represent the information contained in a time series, not as a function of time but as a function of period (or frequency). This representation, where the presence of a specific periodic wave appears as a peak, is called a "spectrum." Using a Fourier transform is a classic – though not fail-safe – method of obtaining such spectra.

2 For example, for determining the periods of lunar or solar tides over series of several years, or for the harmonics and fundamentals of a sustained musical tone.

3 The important factor here is the small number of points, only about ten; we must not be over-awed by the fact that the series represents 300 Ma.

such a periodicity, which disappeared if one arbitrarily introduced a new boundary within a rather long stage of the Cretaceous. So the periodicity might be nothing more than an artifact of this scale. Well and good – except that the scale was evolved precisely on the basis of the principal extinctions, using them to define its stages. If the periodicity of the extinctions is indeed real, it would hardly be surprising to find it reflected in the scale! In the wake of the substantial growth in the 1990s of our knowledge of the precise age of the traps, Rampino and Stothers now differ. The former still upholds the idea of an approximate, though not rigorous, periodicity on the order of 30 Ma. But the latter now believes he can firmly state that the sequence is clearly not periodic. Rampino and Caldeira believe this mysterious periodicity is reflected not only in the extinctions and traps but also in the fluctuations in sea level, the episodes of mountain-chain formation, the abrupt variations in sea-floor spreading, the climatic events that led to the formation of black shales (from anoxic events) or evaporites (from salt) in sediments, and more. Could this cyclic pattern be one of the signatures, a "pulse," of the Earth's dynamics?

Nemesis, the death star

In 1984, Walter Alvarez and his colleague Richard Muller, a Berkeley astronomer, analyzed about 15 impact craters dated at less than 250 Ma. Amid a good deal of background noise, their "spectrum" (see Note 1) seemed to show a peak corresponding to a periodicity of 28 Ma. This figure from their article in *Nature* by itself gave grounds for pause, and the data on which it was based seemed pretty scant and rather fragile. Yet it still would serve to launch a new and much-discussed theory. Richard Muller and his associates, in fact, claimed the famous periodicity was evidence that the Sun has a companion, which they called Nemesis: in effect, our solar system would be a double star. They conjectured that this small companion, in a very elongated elliptical orbit around the Sun, would pass every 28 Ma near the "Oort cloud." Located far beyond the outermost planets, the Oort cloud is the (hypothetical, but very probable) home of the comets that episodically penetrate through the ranks of the planets, sometimes approaching Earth. Gravitational perturbations caused by the proximity of the new star would periodically increase the number of comet heads ejected toward the interior of the solar system and thus likewise increase the probability that some of these would collide with our planet. This hypothetical star, thought to sow death on Earth every 28 Ma, took its name Nemesis[4] from the Greek goddess of vengeance; but she was also the goddess of distributive justice and of the rhythm of fate, which causes ill fortune to succeed periods of excessive prosperity. The US National Science Foundation has financed a vast program to explore the sky and find the culprit. The reader will not be overly startled to learn that this program has not yet produced a result. No Nemesis at the frontiers of heaven.

The series of crater ages has recently been restudied by Grieve, and, as measurements have improved, the idea of a periodicity of about 30 Ma seems to have evaporated. Meanwhile Stothers, after analyzing seven craters that he believes sufficiently well dated (all more than 5 km in diameter and less than 70 Ma old) likewise found no confirmation for periodicity. However, he does believe that he has observed a good correlation with six boundaries of geological stages. I cannot share his position. Five of these boundaries in fact are minor events of the Cenozoic. Two major events, those of the Middle Miocene and the Upper Eocene, are not associated with any crater. So the only object that might match up with a major event is the Manson Crater in Iowa. The age of this small crater was estimated, in 1989, at 66 Ma, corresponding to the KT boundary. It was immediately claimed as a trace of one of the fragments from the Alvarezes' meteorite. Too small, at 35 km in diameter, to have caused the end of the dinosaurs by itself, it has also lately been redated, more reliably and precisely: it is 74 Ma old. The magnetic polarity of the rocks that melted at the time of the impact was moreover incompatible with the reversed polarity that prevailed at the KT boundary. Exit Manson.

Impacts and reversals

Trusting in the validity of the correlations between impact frequency, extinctions, and climate, some researchers have gone so far as to

4 See David Raup's short and excellent book
The Nemesis Affair, New York, Norton, 1986.

argue that impacts might not just modify climates but also trigger glaciation (of which, however, there is not the least geological trace at the end of the Cretaceous). This, in turn, would have modified the position of the axes of inertia of the Earth, entailing an overall drift of the Earth's crust relative to its axis of rotation. This shifting would then have modified the movements in the liquid core and caused reversals of the Earth's magnetic field. In support of this – highly speculative – model, the authors cited as recent examples the parallelism between the age of the Australian and Asian tektites, which cover one-tenth of the Earth's surface and undoubtedly mark the impact of a rather large asteroid, and the age of the last field reversal, 780,000 years ago, that led from the reversed Matuyama magnetic chron to the current "normal" Brunhes chron. They see another match between the tektites of the Ivory Coast and the next-to-last reversal, called the Jaramillo reversal, 970,000 years ago. Precise studies of sediments from high-resolution deep-sea cores (i.e., primarily cores with very high sedimentation rates) have refuted these correspondences. The impact responsible for the Australian tektites occurred 12,000 years before the Brunhes–Matuyama boundary, and that of the Ivory Coast 8000 years after Jaramillo. In the former, the evolution of the atmosphere has been tracked, using oxygen isotopes trapped in ice cores sampled in Antarctica, and it was not possible to establish any correlation, or any cause–effect relation. What's more, no mass extinction has been noted in conjunction with these reversals.

Though I have taught a university course in signal processing for several years, I willingly admit my perplexity at the exchanges of arguments about the "final" demonstration or refutation of all these correlations by eminent specialists. Surely it is not really reasonable to apply such sophisticated methods to such short temporal series, cases in which they can never offer anything but ambiguous answers. The message to students is clear: keep your wits about you, and don't be (too) impressed by your professors' categorical pronouncements but do still be a little impressed, all the same . . . The correlation established in the last chapter (Fig. 6.1, p. 98) does not seem to me to show any periodicity. The reader may judge.

Did impacts cause the traps?

Nonetheless, the principal extinctions do seem to coincide with the eruption of traps, and some extinctions (but only a very few) perhaps coincide with impacts. So an idea comes to mind, particularly for the KT boundary: couldn't a giant impact have fractured the Earth's crust sufficiently to trigger volcanism? In 1987, Michael Rampino, building on calculations by Tom Ahrens, suggested that a body 10 km in diameter falling to Earth at a speed of more than 10 km per second would dig a crater more than 20 km, and possibly even 40 km, deep.[5] Rampino suggested that its traces should be sought under the Deccan Traps. Although our direct knowledge from drilling the basement under the traps is almost nil, the formations formerly covered by the lava and now exposed by erosion show neither fractures nor direct traces of shocks that might indicate such an impact. Some authors have gone so far as to compare terrestrial traps with the "seas" on the Moon. We know that the latter do indeed correspond to immense craters, created by gigantic impacts more than 3.5 billion years ago and then filled with lava (see Fig. 2.5 p. 34). But these involved far larger bodies than the one the Alvarezes cited for the KT boundary; the probability of their colliding with Earth, once the first billion years of existence of the solar system were past, is far less than 1 per 100 Ma (or even, possibly, per billion years).

The idea that the impact of an asteroid measuring some tens of kilometers in diameter might quickly lead to the eruption of traps seems to me to be based on a false notion of the Earth's interior – namely the idea, current in the nineteenth century and still taught by some science books, that the Earth's mantle is molten under the solid crust and may overflow almost instantaneously if cracks allow it to. In fact there is little doubt that the asthenosphere[6] contains only a small fraction (a few parts per thousand) of liquid. Even if an impact had stripped away 20 km of the crust's thickness in a crater 100 km in diameter, it still would not have been able to melt the tens of millions of cubic kilometers of mantle required to produce the Deccan lavas in the necessary short time.

5 This depth is transient and would last only a short time. The crater would then return to a final, "static" depth that would be substantially less.

6 The less viscous part of the upper mantle upon which the more rigid lithosphere plates drift.

Other authors have proposed that the impact site should be sought on the opposite side of the Earth from the traps. The seismic energy released by the collision, reconcentrated by a lens effect at a point opposite the initial shock, would cause the rock of the mantle to melt and to be extracted through the cracks produced by the focusing of tensional seismic waves. This mechanism offers an explanation of some deformations observed on the Moon and on Mercury, opposite certain very large basins (or "seas") created by gigantic impacts more than 4 billion years ago. The traces can still be observed in these two heavenly bodies because they have neither erosion nor plate tectonics. The point opposite the Deccan Traps 65 Ma ago was located off the West Coast of North America, on the little Farallon plate that has since then been subducted and redigested by the mantle. So there is no hope of finding the least trace of it.[7]

This said, a decisive argument against the impact–volcano link, so far as the KT boundary is concerned, is that the volcanism started in a period when the polarity of the Earth's magnetic field was normal. This is incompatible with the fact that the impact signatures (iridium and shocked quartz) are found amid sediments with a reversed polarity.

Where do the plumes come from?

If the plumes cannot be triggered by impacts, what then is the origin of these enormous and rare instabilities in the mantle? The fixed position of the hot spots relative to one another over several tens of millions of years, and their relative independence from the overall course of continental drift, have led Jason Morgan to suggest that they must originate at the base of the lower mantle, not far from the Earth's core. This idea seems to run up against several difficulties: at a depth of 450–670 km, in what seismologists call the transition zone, the velocity of seismic waves increases rather rapidly. Under the action of growing pressure, the principal mineral of the mantle, olivine,[8] is transformed into more compact, denser species called

---

7 And we will see in the next chapter that the point opposite the so-called Chicxulub crater was located 3000 km off the coast of India and could not coincide with the Deccan Traps. So the focusing hypothesis does not seem to work (at least in that case). (See also Fig. 1.4, p. 19.)

8 A ferro-magnesian silicate with the general composition $(FeMg)_2 SiO_4$.

"spinels" and "perovskites." Viscosity undoubtedly rises rather sharply in these, and the convection rate in the lower mantle must thereby be substantially reduced, compared with the typical velocities in the upper mantle (which range from 1 to 10 cm per year). Some think that plumes originating from the lower mantle would have a hard time crossing the mechanical barrier of the transition zone. Geochemists, particularly Claude Allègre and his associates, have shown that the basalts emitted along the oceanic ridges came from a reservoir of a different chemical nature than those poured out on the ocean islands that mark the path of hot spots. The part of the mantle that is "sampled" via the oceanic ridges is depleted in a certain number of elements, because the continental crust was extracted from it 3 billion years ago. Oceanic islands, however, draw upon a deeper mantle. So Claude Allègre would rather locate the source of the hot spots in the transition zone that separates the upper and lower mantle. Yet he accepts the idea that some might come from the lower mantle.

Other researchers, myself among them, think that if this transition zone is not crossed (but is it really impossible to cross?), a deep instability, presenting a considerable anomaly in temperature and density, might in turn trigger a "second-generation" instability in the upper mantle (see Fig. 5.5, p. 87). Finally, still others, like Don Anderson, go so far as to imagine, partly perhaps to generate some discussion, that the hot spots do not have a deep source but form at the base of the lithosphere in contact with the heterogeneous zones of the parts underlying it. Seismic tomography as yet offers us no way to resolve the controversy, although Henri-Claude Nataf, of the Ecole normale supérieure de Paris, thinks he has detected weak seismic propagation anomalies at the base of the lower mantle under the Pacific, where he believes the base of a deep plume is located.

Thus the partisans of "single-layer" and "two-layer" convection in the mantle have been arguing for some time. Each of the camps can adduce a considerable amount of convincing evidence against the other. Recent calculations, on more and more powerful computers, may show a way out of this dilemma. Philippe Machetel,[9] then at the

9 See Philippe Machetel, La convection dans le manteau terrestre, La Recherche, Paris, 21, 1238–1246, 1990.

University of Toulouse, and Paul Tackley, then at Caltech, showed that the transition zone does indeed constitute a barrier to the passage of matter from one part of the mantle to the other, but only in "normal" times. The masses of cold matter contributed by plates undergoing subduction may accumulate on top of the transition zone and thus grow into an abnormally heavy package, which might all at once (in geological terms) sink into the lower mantle.[10] Likewise, an accumulation of hot, lightweight material originating from the lower mantle might episodically erupt into the upper mantle. This intermittent convection might well reconcile the adherents of normal dynamics (the longest phases) with those who believe that abrupt, episodic phenomena are possible. It is some of these events that perhaps may be linked with the mass extinctions.

If the plumes that generate the traps come from the base of the upper mantle, it does not seem likely, by analogy with the experiments described above, that the diameter of their heads could exceed 200 km initially, or approximately 500 km after they spread out under the lithosphere. The volume of abnormally hot mantle needed to produce the quantity of basalt in the great traps is such that an origin at the base of the mantle seems plausible to me. Calculations show that the head of a thermal plume, which entrains and heats supplementary material from the nearby mantle as it rises, reaches a diameter of 1000 km and may then spread across more than 2000 km at the base of the lithosphere, which is in good agreement with the lateral extent of the major traps. Let us note that the total amount of heat currently being transported by the few dozen active plumes is estimated at less than 10 percent of what the Earth loses from its surface.[11] The heat flux that reaches the base of the boundary layers represented by the plates is in a sense the source of energy for their drift, while the heat lost by the core is partly the energy source for

---

10 Seismologists have now improved the resolution of seismic tomography to the point where they can see the remnants of cold subducted plates falling into the lower mantle.

11 This quantity is on the order of 80 milliwatts per m², the equivalent of one 100 watt light bulb per 1250 m². The heat lost by the Earth originates from a number of different phenomena: dissipation of the primordial heat associated with

the gigantic impacts at the time of formation of the planet; latent heat released from the core by crystallization at the surface of its central, solid part (the inner core); heat corresponding to the dissipation of gravitational energy; and finally, heat produced by the decay of the radioactive isotopes of uranium, thorium, and potassium, primarily concentrated in the crust.

the plumes. The very different forms that convection takes result from pressure and temperature conditions, and especially from the laws of rock flow, which vary enormously with temperature. The rigid, cooled plates, as they return to the mantle by subduction, cause a flow geometry that is totally distinct, and largely independent, from that induced by the cylindrical columns of the plumes. Plates and plumes are, therefore, two complementary aspects of the mantle's dynamics. Their very different manifestations at the surface undoubtedly explain why the founders of plate tectonics – except for Jason Morgan – underestimated the significance of plumes.

Plumes and reversals

A totally independent observation may possibly allow us to link the way in which the core functions to the way the mantle functions, and thus provide support for the idea of a very deep source for plumes. But its interpretation, as we will now see, is still very controversial. The observation goes back to 1972 and was made by Peter Vogt. After successive intervals of dormancy or incredulous dismissal, it was picked up again in the mid-1980s by McFadden and Merrill, by Loper and his associates, and also by Jean Besse and myself. It has a bearing on the evolution of magnetic field reversals over time. Some observers, among them myself, suggest that the very long-term variations in the frequency of these reversals might be linked with the two greatest mass extinctions.

We have seen how, notably with the work of Lowrie and Alvarez at Gubbio, a scale for reversals of the Earth's magnetic field has little by little been developed. This scale is fairly well understood for the past 160 Ma, a period during which the magnetic anomalies are "painted" onto the ocean floors. It is less clear for previous periods, for which our knowledge must be based on magnetostratigraphic measurements in outcropping sedimentary sections (see Chapter 2). Viewed on a scale of several Ma, polarity reversals seem to occur randomly (with a mean frequency of the order of four reversals per Ma for the recent period); but if we look at longer periods we realize that this mean frequency varies considerably (Fig. 7.1). It has increased, more or less regularly, for the past 85 Ma.

In detail, the curve for the frequency of field reversals seems to

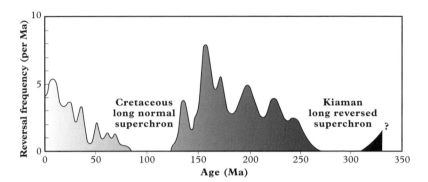

Figure 7.1
Changes in magnetic reversal frequency in the last 300 Ma. Two unusually long periods without any reversal are indicated as superchrons.

show fluctuations that may look periodic. Like the periodicity of extinctions or impact craters, these fluctuations are still vigorously debated. The reversal sequence, however, is the most detailed and the most reliable of all the time series studied so far. Relatively higher-frequency periods occurred 10 and 40 Ma ago, but also 25 Ma ago and, with less certainty, 55 and 70 Ma ago. In 1983, Alain Mazaud and Carlo Laj proposed a periodicity in the order of 15 Ma. Stothers has argued for one of 30 Ma, while the statisticians Lutz and McFadden see nothing but artifacts, fluctuations of a statistically random nature. Loper and McCartney have compared these quasi-periodic fluctuations with those discovered in other time series, particularly the extinctions, and believe they have found signs of a supplementary correlation between these events in the biosphere and phenomena that can only have their source in the core. Nevertheless, doubt remains as to whether this periodicity even exists.

Around 85 Ma ago, for an exceptionally long time of about 35 Ma, the polarity of the Earth's magnetic field remained locked in the "normal" direction it has now. This is the long interval (or superchron) of the Cretaceous. Before this period, the frequency of field reversals had declined, starting from a maximum of the same order as the current value. Going farther back in the past, we find an exceptional period of immobility, this time with a reversed polarity, that lasted 70 Ma, during the Carboniferous and Permian Periods at the end of the Paleozoic. This has been named the Kiaman superchron by the great paleomagnetician Ted Irving, who established its importance as long ago as the late 1950s.[12]

For even more ancient times, the data are very rare and spotty. The existence of a third period of calm during the Ordovician has been suggested; it is currently still under research. So the evolution of the frequency of magnetic reversals over time seems to be modulated in accordance with a pseudo-period, a characteristic time constant, on the order of 200 Ma. It is hard not to be impressed when one finds that two of the greatest traps, those of the Deccan and Siberia, which are associated with the two greatest biological crises the Earth has known in more than 300 Ma, both happened shortly after the two exceptional phases of "magnetic immobility" – the Cretaceous and Kiaman superchrons.

The terrestrial dynamo

If we assume it is no coincidence that major mass extinction events seem to correspond with the surface emergence of convective plumes originating deep in the mantle, what mechanism could link the latter to the reversals of the magnetic field? The major portion of the Earth's magnetic field results from the existence of electric currents in the Earth's iron core. Left alone, such currents would die out under the action of the Joule effect, dissipating heat, in less than 10,000 years. Now, paleomagnetism tells us that a magnetic field has existed for more than 3 billion years, and undoubtedly almost since the origin of our planet.[13] So there must be some mechanism that sustains this field. This mechanism is linked to the existence of powerful, rapid convection movements within the outer, fluid part of the core and is related to the mechanism by which a dynamo can function. Great names have been associated with the difficult task of applying the dynamo theory to the origination of the magnetic field in the Earth's core. Among the pioneers, we can mention Walter Elsasser and Teddy Bullard.

---

12 Kiama is the name of an Australian village where a forerunner, the physicist Chevalier, established back in 1925 that very ancient rocks had been magnetized in a direction opposite to the present one – this more than 30 years before the reality of such field reversals had been commonly accepted.

13 Or at the very least since the core formed, only 50 Ma into an existence that (as the reader will recall) goes back 4.5 billion years.

These fluid movements of the core are reflected on the Earth's surface by the secular change in the magnetic field, and data recorded at observatories over less than three centuries indicate that the velocity involved is a few kilometers or tens of kilometers per year. This may seem like a rather low figure, but such currents are in fact 100,000 times faster than those that move the plates of the lithosphere. In the presence of an initial magnetic field, these movements are able to induce an electric field: this is Faraday's law. This field creates electric currents within the conducting fluid, and these in turn generate a magnetic field. This new field is added to the one that originally existed and may reinforce it. If the movement is vigorous enough, or the geometry of the convection is efficient enough, the original field becomes unnecessary and the magnetic field itself then continuously induces the electric fields that sustain it. In this case we speak of a "self-excited" (or self-sustaining) dynamo. A simple example (Fig. 7.2a) is a conducting disk rotating around its axis in a constant magnetic field to which it is parallel. The rotation induces an electric field in the disk, and if the rim of the disk is attached to the axis of rotation by a wire, an electric current will flow. Then all we have to do is form the wire into a loop parallel to the disk and make the device turn fast enough, and the original magnetic field will become unnecessary.

This dynamo does not violate the laws of thermodynamics. No perpetual motion is involved. In the case of the disk dynamo, the energy is contributed to the system by rotation. The stirring movements within the core have several possible sources of energy: heat imprisoned when the core was formed, heat released by the radioactive elements it contains, the considerable heat released by the crystallization of iron at the surface of the inner core (see Note 11), and finally lighter and gravitationally unstable elements released into the liquid at the time of this crystallization. This solid inner core grows very slowly at the expense of the liquid part. After billions more years of cooling it will eventually occupy the entire core; movement will then no longer be possible, and the magnetic field will be extinguished. This is why the Moon with its small, solid core no longer has its own magnetic field. The study of the terrestrial dynamo is a vast and complex field of mathematical geophysics, and five decades of effort have taken us only the first few steps down this road.

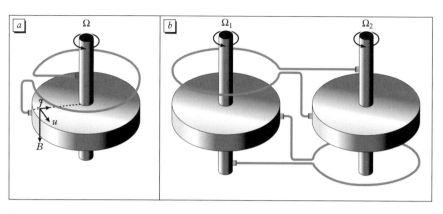

Figure 7.2
Models of simple laboratory disk dynamos (after Rikitake). (*a*) One single disk and current loop generating a magnetic field; (*b*) two coupled disks with the current loop from one acting on the other. The disks are rotated at velocities Ω; at one point, local velocity u, magnetic induction B and resulting induced current $\mathcal{J}$ are shown.

Although the magnetic field produced in the small model of the self-excited disk dynamo never reverses polarity, a system with spontaneous reversals can be constructed by coupling two disk dynamos (Fig. 7.2*b*). In 1958, the Japanese geophysicist Rikitake imagined inducing the magnetic field that surrounds one of the disks through the coil that conducts the current induced in the other disk. Several systems have thus been constructed that generate random reversals whose sequence is reminiscent of that observed in the Earth. By now these have become classic examples of deterministic chaos theory. However, this dynamo, with complex topology but simple movement, bears no great relation to Earth's core, where the topology is simple but the movements may be highly complex.

Gary Glatzmeier, from Los Alamos, and Paul Roberts, from UCLA, have recently produced a complex computer model of the Earth's dynamo that seems to generate a fluctuating, reversing magnetic field that in some ways is reminiscent of the Earth's actual field. Although this is a significant step forward, the physical parameters they must use to make the computation feasible (especially the assumption of a fluid core with a high viscosity) are still remote from the proper values.

The D″ layer

In this manner, researchers are beginning to construct mathematical models of reversing dynamos, but we are still far from a satisfactory understanding of what happens during a reversal, much less an explanation of the very-long-term fluctuations in the frequency of these reversals. However, we do know that the heat lost by the core at the base of the mantle is the principal energy source of the dynamo. Now, at the base of the mantle there is a distinct boundary layer (called the D″ layer[14]) with very special properties that has long been known to seismologists. Extremely heterogeneous, apparently less dense and less viscous than the part of the mantle that covers it, the D″ region is about 100 km thick. Knowing the temperature of solidification of iron at the surface of the inner core, and its distribution in the convective liquid core,[15] we can determine the temperature at the base of the D″ layer "from the bottom up." The temperature at its upper surface is determined "from the top down" by the surface temperature and the distribution, again adiabatic, in the convecting part of the mantle.[16] Although many uncertainties remain, particularly regarding the thermal nature of the transition zone, we can estimate that the temperature jump across the D″ layer, which is on the order of 1000°C, remains substantially constant over time.

The heat flux extracted from the core and transmitted by conduction across the D″ layer is proportional to this jump in temperature and inversely proportional to the layer's thickness. The heat transmitted by the core initially causes the D″ layer to thicken. But having become less dense and less viscous than the overlying material, and therefore unstable, beyond a certain critical thickness this layer emits plumes. Then its thickness decreases. Thus, it is possible that the system may oscillate between periods of repose and periods of instability when plumes are emitted. The resulting variations in

14 The earlier letters of the alphabet were assigned to the more superficial layers, but today this nomenclature has gone out of use, and only the D″ layer has survived in the terminology.
15 In the case of convection within the core, the movement is rapid enough that a small amount of fluid does not have time to reach equilibrium with its surrounding environment. This causes a temperature gradient, called an adiabatic gradi-

ent. Any thermodynamic transformation without an exchange of heat is called adiabatic. In the parts of the core and mantle involved in convection movements, the adiabatic gradient is several tenths of a degree Kelvin or Celsius per kilometer.
16 See Jean-Paul Poirier, *Les profondeurs de la terre*, Paris, Masson, 1991.

thickness of the D″ layer would entail variations in the heat flux extracted from the core and, consequently, the production of instabilities that would trigger reversals of the magnetic field in the core. Let us mention right away that some authors have constructed analogous models in which the sign of the correlation is just the opposite: the frequency of reversals increases with the thickness of the D″ layer. Without going into too much detail, this shows that the problem is far from a satisfactory solution. It offers an opening for potentially fruitful research. It may be possible to link phenomena as deep as reversals of the magnetic field to phenomena as superficial as the climatic alterations resulting from the eruption of traps. The natural link between the two would lie in the existence of deep plumes from the mantle, and the manner in which they function.

From the core to the biosphere: the missing link

According to Loper, the fluctuations every 15 Ma in the frequency of field reversals would, therefore, reflect cycles of instability in the D″ layer and the concomitant emission of plumes. Though it is founded on only two examples, I find the correlation between the two abnormally long periods of stability when the field did not reverse polarity and the two greatest extinctions in the past 300 Ma, which mark the end of the Paleozoic and Cenozoic Eras, fascinating. In both cases, the D″ layer may well have attained an entirely abnormal thickness, and gigantic diapirs (see Fig. 5.3, p. 79) would finally have torn away from it. The reversals of the magnetic field would immediately have resumed, while the plumes, continuing their ascent through the mantle at a velocity of several tens of centimeters per year, would then either have crossed the transition zone themselves or have triggered "second-generation" plumes there. Arriving in the upper mantle, the plumes would finally have reached the base of the plates causing the lithosphere to bulge and thin out for several Ma. They would finally have triggered the eruption of the traps, about 15 Ma after the plumes started rising (in Siberia 250 Ma ago and in India 65 Ma ago).

This model is, of course, speculative. It does not explain why the variations in the frequency of field reversals are far fewer in the case of the other traps, whose volume nevertheless does not seem a great deal smaller (except in the case of the Columbia Plateau). Roger Larson has even proposed an opposite correlation. A marine geophysicist specializing in the study of ocean floors, Larson attributes to the plumes the formation of enormous plateaus of abnormal oceanic crust that stud the floor of one part of the Pacific. The largest of these is the Ontong–Java Plateau, in the western Pacific (see Fig. 5.4, p. 82), whose volume he estimates at 50 million km^3. From a compilation of the age and volume of some 25 submarine plateaus, Larson has calculated that the volcanic production that created them underwent a sudden increase about 120 Ma ago, at the time of the eruption of the Ontong–Java Plateau. According to him, this event marks the arrival at the surface of a superplume, which almost instantaneously resulted in the cessation of reversals. This assumes that the plume rose at a velocity of several meters per year, that the D″ layer very suddenly grew thinner (as a result of the departure of the plumes), and that the resulting increase in heat flux would have to inhibit the reversal process instead of exciting it. The very rapid rate of ascent is all the more surprising since laboratory experiments show that a new plume takes rather a long time to push its way through the mantle if the latter has not been heated already.

Under Larson's hypothesis only one event of this type has happened since the Paleozoic (at least 300 Ma). Moreover, it would have to be entirely without relation to the mass extinctions: for, as we have seen, there was no first-order event at the time of the submarine eruption of Ontong–Java. Some authors, such as Mark Richards, have pointed out that the volumes of the oceanic plateaus might lead one to overestimate their importance in relation to the continental traps, in which, we must recall, three-quarters of the magma doubtless remains underplated at the base of the lithosphere and does not appear in the calculated totals. Other authors, like Anny Cazenave of Toulouse, think the Ontong–Java volcanism does not correspond to the same plume process as the traps. Whatever the case may be, the reader can see it's a long way from an alluring suggestion to a quantitative model.

Figure 7.3
Shiva sculpture in the Ajanta Temple (India). Basalt (the very lava that may have killed the
dinosaurs). (Photograph by J. Schotsmans.)

Nemesis or Shiva?

Stephen Jay Gould has remarked that surely it was inappropriate to
propose the name Nemesis for Muller's death star. Nemesis was the
goddess of calculated vengeance. Yet what could be more unexpected, less merited, more contingent than this catastrophe that
closes the Mesozoic and then goes on to permit this astonishing
resumption of Life, leading evolution into a kind of "experimenta-
tion run riot" at the start of the Tertiary? Gould thought it more

appropriate to name the star after Shiva, a Hindu deity associated, like Nemesis, with destruction but also with rebirth. Intrigued by this image, a few years ago I asked Jeanine Schotsmans, a curator at the Brussels museum, to lend me one of her superb photographs of a Shiva sculpted on the walls of the Temple of Ajanta in India, in the very lava that I believe may have been at least partly responsible for the massacre of the dinosaurs (Fig. 7.3). I suggested to the editors of *Nature* that they should put this photograph on the cover of the issue in which our first argon datings of the traps were published.[17] An accompanying caption (with a wink to Stephen Gould) was unfortunately omitted by the journal's editors. A good many readers of *Nature* must have wondered what this divinity was doing on the cover of their magazine.

The correlation established in the preceding chapter is quite a trophy for the "volcanists" to add to their hunting bag. Seven traps coincide with seven periods of extinction. The boundaries of the three great geological eras, those first-order interruptions observed in Nature since the eighteenth century, are also among the seven. Must the fruitful impact hypothesis then join the hundred or so hypotheses, involving greater or lesser amounts of fantasy, that have had to be abandoned one after another over the past 100 years? But that would be getting ahead of ourselves . . .

17 In the same issue, Duncan and Pyle published important companion results that confirmed the brevity of the eruptions and their correspondence with the KT boundary.

8

Chicxulub

Most of the craters that might have marked the meteorite's point of impact, such as the Manson Crater, turned out to be too small or the wrong age. Granted, a quarter of the ocean floor that existed at the time has disappeared into the subduction zones, and the crater may very well have been dragged down into the mantle with them. But the shocked quartz discovered in many sections of the KT boundary suggested an impact on the continental crust instead,[1] and the "impactists" strongly hoped that they would ultimately discover the "smoking gun." So they looked for the crater, the trace of the gigantic impact that would prove their theory's validity.

The hunt for the crater

The abundance and size of the grains of shocked quartz led them to look for a site not far from North America, possibly in the sea.[2] According to Tom Ahrens's calculations, the impact of an asteroid 10 km in diameter in the open ocean would have caused a gigantic tidal wave: initially well over 1 km high(!) it would then have subsided in proportion to the distance from the impact point, tearing up rock from the bottom of the ocean itself, and devastating and eroding the continental plateaus and near coastal zones. Some layers of sediment sampled by submarine drilling in the Caribbean have been interpreted as the deposits from this gigantic tsunami.[3] In Cuba, in fact, a layer ranging from 5 to 450 m thick and containing at its base blocks more than 1 m in diameter was interpreted as an immense

---

1 The continental crust is characterized by granitic rocks with a rather high quartz content. The oceanic crust, basaltic in nature, has practically no such quartz at all.
2 The continental crust also extends under the sea, forming shallow platforms. The oceanic crust generally appears in deeper basins, the abyssal plains, the mid-ocean ridges, and the oceanic trenches. (It may also be covered by a thicker or thinner layer of sediment.)
3 This word, of Japanese origin, actually refers to tidal waves triggered by earthquakes.

"turbidite,"[4] the result of an abrupt uprooting and rapid deposition of materials in the immediate vicinity of the impact.

A typical section of the KT boundary in North America starts with a layer of clay 2 cm thick, which the meteorite's supporters interpret as a deposit of ejecta, the least energetic elements dispersed by the impact; this clay layer is succeeded by the "worldwide" layer, a few millimeters thick, where iridium, shocked quartz, spinels and spherules, soot, and isotopic anomalies – the results of the vaporization of the projectile and part of its target – are concentrated. But on the island of Haiti, near Béloc, there is a substantial layer, 50 cm thick, interpreted by some as being of volcanic origin; in 1990 Alan Hildebrand and William Boynton attributed this layer to the impact. It was the thickest layer of ejecta – severely altered, of course – discovered to date.

Reviewing two sites from which cores had been drilled during the international Deep Sea Drilling Project (DSDP) in the southwestern Gulf of Mexico, Walter Alvarez and several of his colleagues believed they recognized the sedimentary sequence caused by the huge wave. According to them, the absence from the strata of the last stages of the Cretaceous Period resulted from the catastrophic erosion the wave had produced. A pebbly clay represented a submarine slide. A cross-bedded sandstone, more than 2 m thick and containing shocked quartz, tektites, and iridium, represented the ejecta picked up in the erosion caused by the wave. I have to emphasize the cautious and conditional tone of this 1992 article, which contrasts with the tone of conviction beyond all appeal that would dominate some scientific works in subsequent years.

Also in 1992, the same authors, led this time by Jan Smit, described a section from northeastern Mexico at Arroyo el Mimbral (Fig. 1.4, p. 19) where the KT boundary corresponded to a layer as much as 7 m thick. From bottom to top, it showed a bed of spherules, again interpreted as redeposited ejecta; massive, lenticular, or laminated sandstones, in which the debris is sorted upward in order of decreasing grain size, interpreted as the sediments washed by the wave from

4 Turbidites are sedimentary deposits, generally from the deep sea, that result from the rapid deposition of sediments with a wide range of grain sizes. These deposits are caused by rather sudden floods of submarine mud, sometimes containing blocks of considerable size, triggered by earthquakes or severe storms.

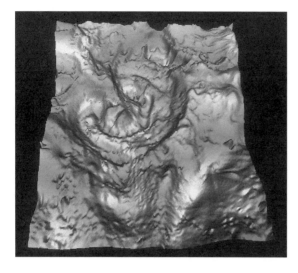

Figure 8.1
The Chicxulub
impact crater in
Yucatàn, revealed by
gravity anomalies.
The multi-ring crater
is about 200 km in
diameter (V.
Sharpton).

the surrounding coastal plains as it retreated; and finally sand beds,
interrupted by layers of clay, interpreted as the cyclic deposits resulting from those oscillations of an oceanic basin known as "seiches."[5]

The discovery of Chicxulub

Cuba, Haiti, Mexico, the Gulf of Mexico: the noose was tightening.
In 1990, Hildebrand and Boynton believed they had finally found the
culprit, in the form of a circular structure nearly 300 km in diameter,
buried under 3 km of sediment in the ocean basin of Colombia. A
short time later, however, they switched suspects. They had dug out
the abstract of a paper presented at a convention a good ten years
before in which two petroleum geologists, Glen Penfield and Antonio
Camargo, had described a subterranean structure on the northwestern edge of Yucatán. Invisible at the surface, it was indirectly
revealed by anomalies in the gravitational and magnetic fields. This
circular structure, revealed primarily by gravimetry (Fig. 8.1), was
almost 200 km in diameter and was reminiscent in shape of other
circular anomalies, such as the one found above the Manicouagan

5 These are damped oscillations similar to those   basin full of water and then let the water return
that develop when you suddenly agitate a big   to equilibrium.

impact crater in Canada. Penfield and Camargo immediately had the idea that this observation might somehow be set in relation to the Alvarezes' recent theory. But their ideas, proposed in 1981, went unnoticed and were left to hibernate for nearly a decade in a collection of abstracts. In fact, the paper had been presented at a conference of the Society of Exploration Geophysicists, whose primary concern was really not the end of the dinosaurs. To be fully appreciated, a discovery must be published at the right time, among the right audience.

Cores had been drilled near the structure in the early 1960s, for petroleum exploration. The drilling closest to its center, not far from the village of Chicxulub (which has now given the structure its name), had encountered layers that were first interpreted as a geologically "normal" sequence of limestones, andesite lava flows,[6] and volcanic ash. But a new analysis revealed grains of shocked quartz. Did this set of cores quite simply represent a geological section of the filling of a crater by breccia that fell back into the hole after the impact, and by rock melted by the released heat (a melt sheet)?

The age of the crater

Attention immediately turned to the sections close to the KT boundary that must have been in the "ringside seats" at the time of the cataclysm and thus must have borne its mark. The spherules from the section near Béloc in Haiti included well-preserved glassy material. Quickly interpreted as melt droplets caused by the impact of Chicxulub, which had just been discovered, they were dated simultaneously by Pierre-Yves Gillot of the University of Orsay and by Glenn Izett and Brent Dalrymple, using variants of the argon isotope method. The ages they found – 64.0 Ma according to the former, 64.5 million according to the latter – had remarkably low uncertainties (0.7 and 0.1 Ma, respectively). These ages were compatible with one another and with the absolute age assumed for the KT boundary (64.3 ± 0.3 Ma, according to the results of Baadsgaard in

6 Typical of the volcanoes of the Pacific Ring of Fire, particularly in the Andes, from which they get their name.

Montana[7]). There is, moreover, no doubt that they are stratigraphically close to this boundary, whose age they will help establish from now on.

In *Science* in August 1992, Carl Swisher and his colleagues published their dating of three tiny fragments of andesite glass, each weighing two-tenths of a milligram, recovered 1390 metres deep in the "Chicxulub 1" drill cores. The results indicated an extraordinarily precise age of 64.98 ± 0.05 Ma. In the same study, Swisher found an age of 65.01 ± 0.08 Ma for the Haitian tektites. These results are almost too good to be accepted without question (see note 7); nevertheless they show how much progress has been made with argon dating applied to tiny grains heated with a laser.

If the asteroid did in fact hit northern Yucatán, the impact occurred on the continental crust, and in shallow water. It could not have generated a tsunami as gigantic as the one Ahrens had originally imagined – the water was too shallow. But the dissipated energy would correspond to an earthquake with a magnitude greater than 10 (and according to some, even 12) on the Richter scale,[8] and could very readily have triggered great earth slides within a vast radius. These would explain the observations in the Mimbral and Béloc sections. In a series of articles, Sigurdsson suggests that the asteroid produced considerable quantities of dioxides of carbon and sulfur, because the rocks of the Yucatán "bull's eye" include carbonates (limestone) and sulfates (gypsum and anhydrite), which could have been chemically broken down by the shock.

I must again refer to Tom Krogh's very fine study (see Chapter 2). You may remember that he applied the uranium/lead isotope

7 This result was obtained with sections from the USA and Canada using three different methods. The experimental uncertainty here is meaningful only in relative terms. If we want to compare these results with those obtained with other specimens in other laboratories, we must remember that the neutron flux used to irradiate the specimens is not known with a precision of more than 1 percent, and that the ages used for the standard specimens are not exactly the same in all laboratories. So the absolute age of the KT boundary is known only with an uncertainty of 1 Ma. Throughout this book (or almost, anyway), I have rounded it off to 65 Ma.

8 The magnitude of an earthquake is correlated with its total energy. Its measurement is not unique, and in fact it is common to define several magnitudes for a single earthquake. The scale in general use, called the Richter scale, is logarithmic. The greatest historic earthquakes have seldom attained 9 on this scale. A magnitude 10 earthquake would be about 50 times more powerful than a magnitude 9 quake. An earthquake with a magnitude of 12 (if such a thing has ever happened) would have an energy 6 million times greater than the great San Francisco earthquake of 1906!

method to tiny grains of zircon sampled from the KT boundary in a variety of North American sections. Krogh discovered that these specimens preserved the record of two ages: the age of the ancient crust to which they belonged, on the order of 400 to 500 Ma, and the age of an event that had perturbed them at the KT boundary. The older age is that of the basement near Chicxulub, and not that of the basement near the sections where they were sampled. So the crust of Chicxulub may well have been the source of these particles, which were then transported far from the impact site.

The greatest impact in the solar system?

So the matter seems settled. Caution gives way to certainty. To see for yourself, you need only read the latest articles by Virgil Sharpton. According to this author, the andesite glass does indeed represent the melt sheet caused by the impact. It contains iridium, it is the right age, the crater was large enough that there is no need to resort to multiple impacts or showers of comets. Even the magnetic polarity – reversed – is right. So there is no doubt: the impact site of the Alvarez asteroid has been found!

The time has come to "dissect" the Chicxulub crater. A new analysis of the gravimetric data reveals an impact basin with multiple rings (between two and four concentric rings depending upon who interprets the data) with an outer diameter of 200 to 300 km. Chicxulub has been compared with multi-ringed impact craters on other planets; in September 1993, Sharpton concluded that it was one of the largest impact structures to have occurred in the part of the solar system within the asteroid belt, that is from Mercury to Mars, since the great period of bombardment ended, nearly 4 billion years ago.

The tone is just as final in an article that Robert Rocchia published in *La Recherche* in December 1993. We need only look at a few subheads[9] and a sentence from the abstract: "Only the impact of an extraterrestrial object can generate lamellar defects in impact quartz." "The results obtained from the minerals mark the end of a

9 Though we must never forget that in a journal article the subheads are chosen by the editors, sometimes to the authors' great dismay.

controversy." "The recent results obtained from the minerals by the laboratories at Gif and Lille . . . support the extraterrestrial scenario beyond a doubt."

So what are these results? We have touched upon them in Chapter 2 (Fig. 2.7, p. 40). Jean-Claude Doukhan and his team, early in the 1980s, applied high-resolution transmission electron microscopy to give a detailed description, identification, and interpretation of the very peculiar defects in shocked quartz that had first been observed with far less precision under the optical microscope. Crystalline twins and parallel lamellae of glass, measuring tenths of a micrometer, are found in the grains from the KT limit and in specimens from confirmed impact sites. No specimen of indisputable volcanic origin observed to date has included them. Laboratory experiments show that a pressure reaching the values generated on impact, but statically – maintained long-term and not briefly as in the case of a shock – does not produce these highly characteristic microstructures.

We owe another observation to Jan Smit and Francis Kyte: magnetites containing nickel, preserved almost intact in some spherules, have been interpreted as remnants of meteoritic material. These magnetites do not originally exist in the meteorite. They may form on the surface of its fragments when these are heated and melt as they fall through the atmosphere (or when they recrystallize from the material vaporized on impact). According to Robert Rocchia and his colleagues, the high level of oxidation of the iron in these magnetites, and their high nickel content, can only be explained by oxidation in the atmosphere of high-nickel meteoric material (1 percent nickel). Finally, as we have seen, these magnetites are distributed across a stratigraphic thickness far thinner than that of iridium, which is thought to have diffused into the rocks after its deposition. The highly variable composition of the magnetites, however, has led Robert Rocchia to doubt the existence of a single impact site, particularly one at Chicxulub. In fact, he observes the highest concentration of what he interprets as pure meteorite debris not in Yucatán but in the Pacific. Walter Alvarez has shown how the apparently awkward distribution of some of the ejecta could result from the combined effects of the impact angle of ejection and the Earth's rotation.

Doubts

In the scheme of good professional practice universally accepted (at least in theory) by researchers, a result can only be cited and discussed after it has been reviewed and approved by "referees" and published. However, I would like to cite several results gleaned from abstracts of conferences and some unpublished articles (what scientists call the "gray literature"), which provide food for some doubt. Let's follow these trails.

Going back to the analysis of the layer of ejecta in Cuba, the Big Boulder Bed, a Cuban geologist believes that these blocks result from the alteration of the bottom part of a very thick turbidite that is otherwise "normal" – i.e., of terrestrial origin. This observation reminds us how important it is to perform a careful geological analysis of these sections, combining field methods, sedimentology, stratigraphy, and tectonics. The specimens from this site would ultimately be analyzed with highly sensitive ultramodern technologies, but by other researchers who often had not the slightest clue of the specimens' *in situ* position.

Based on a new study of the oceanic cores from the Caribbean and Gulf of Mexico, Gerta Keller concludes that the deposits supposedly caused by the tsunami generated by the impact are in fact older than the KT boundary and represent submarine flows and conventional terrestrial turbidites. In fact, in 16 sections from this region, the KT boundary and the several hundred thousand years that surround it are quite simply absent. This gap, which as we know is common at this boundary (as it also is at the Permo–Triassic boundary), prevents our knowing anything at all about what happened at the time. In association with other colleagues, including Wolfgang Stinnesbeck, Gerta Keller then reanalyzed the most complete sections of the region, especially the one at Mimbral. These researchers proposed that the single layer thought to have been deposited in a few instants by the tsunami in fact seemed to contain three distinct units, deposited in channels. These series date from the Upper Cretaceous Period, and the KT boundary is at their top. Many observations (the presence of erosion surfaces; a layer of sand that was already consolidated when the next layer was deposited; a clean contact face between the two lower units, implying that the first was already consolidated when the second was deposited; sedimentation intervals corresponding to different deposition environments; a layer with bioturbation near the top, indicating normal sedimentation and the presence of burrowing organisms just before the deposition of the first Tertiary clays) are cited as proof of a long deposition, not an instantaneous one.

As for the Béloc section, several authors have refuted its impact origin. Lyons and Officer, for example, have emphasized that over 95 percent of the deposits consisted of clayey minerals characteristic of the alteration of volcanic glass. The particles of unaltered glass have seldom preserved their original surface; they are andesitic and filled with vesicles generally not found in tektites. Both C. Jéhanno and Robert Rocchia's group emphasize that the beds of globules are multiple and sorted by size, and that the low level of oxidation of iron and the absence of iridium would demonstrate that these are, in fact, several volcanic layers that were mixed and redeposited. Yet according to these researchers, the distinct layer of clay that overlies them, rich in iridium and spinels, is indeed the trace of the impact, which coincidentally covers volcanic layers that were mistakenly interpreted as being associated with the impact. Coming from supporters of the extraterrestrial hypothesis, these observations surely carried particular weight. For his part, Hugues Leroux, a student of Jean-Claude Doukhan who has worked with Robert Rocchia and Eric Robin, has just found that the distribution of shocked minerals would make it possible to reconcile the spherule bed and the iridium layer: both may be the trace of a single event.

Time to drill again in Chicxulub

The most astonishing remarks on the subject undoubtedly come from E. Lopez Ramos and A. Meyerhoff. Back in 1973, Ramos described the Chicxulub 1 drilling at the center of the circular structure. At a depth of around 1000 m, he shows assemblages of fossils typical of the Upper Cretaceous, ranging from the Campanian to the late Maastrichtian. These fossils come from horizontal layers of compact marls without the least sign of later disturbance. As for Meyerhoff, at the time of the drilling he was a consultant for the Mexican petroleum company Pemex, and in this capacity was closely associated with the biostratigraphic dating of the cores between 1965 and 1977. But where are those cores today?

For some time it was thought that most of the specimens had been destroyed in a fire at the warehouse where they were stored. Meyerhoff reports the observations he delivered on the drilling of Yucatán 6, 30 km from the center of the "crater." Drilled in 1966, this core crossed an ordered sequence of Cenozoic rocks, then 350 metres of Cretaceous rock, and finally a volcanic sequence. The volcanic part shows six successive layers of andesite lavas covered by bentonites (layers of fine ash that indicate a phase when volcanism stopped, followed by alteration). The Cretaceous fossils are above and *between* the flows of the volcanic sequence. How could the impact have left intact such shallow layers, older than itself? It must have vaporized, melted and overturned the crust for a depth of more than 10 km!

In late 1991, it was learned that the missing cores had been found, some of them at the University of Mexico. They were soon reanalyzed.[10] Several groups confirmed the presence of melt rock and dated it precisely to the KT period. The scientific community is now calling for a new round of deep drilling, with full and controlled core sampling, on the structure of Chicxulub. Mexican scientists, for their part, have completed some shallower and less expensive drilling at a few sites on the edges of the crater.

Two abstracts brought out in conferences at the end of 1993 offered further conflicting elements that still require confirmation. Hansen and Toft claimed to have discovered grains of shocked quartz with characteristic families of planar defects in layers of rhyolite ash from the Upper Paleocene Epoch, in Denmark. In Greenland, Nicola Swinburne and other researchers found beds several tens of centimeters thick containing spherules that themselves hold inclusions of reduced iron and spinels with high nickel and iridium contents.[11] These layers, which date from the Paleocene, are associated with the volcanic flows and tuffs of western Greenland. Might these two indicators considered so characteristic of an impact – nickel-bearing magnetites and shocked quartz – be produced by earthly volcanism? This

10 For a lively account of the fate of the cores, see Walter Alvarez, Chapter 2, Note 2.
11 Robert Rocchia believes that they are quite

characteristic of volcanism and unrelated to the spinels he has analyzed elsewhere.

is the North Atlantic province, analogous to the Deccan Traps and evidence of the birth of the Iceland hot spot. Its age matches that of the Paleocene–Eocene boundary. No one so far has suggested that the beds could be explained by an impact. So might it be possible that the terrific explosive eruptions that must accompany the birth of a hot spot are capable of covering the surface with large quantities of iridium, shocked minerals, spinels, and matter with the same composition as the mantle – in short, the sequence of anomalies found at the KT boundary? Though the case for the KT impact was growing stronger, some nagging questions lingered.

Shoot-out at Mimbral

A last major opportunity for an update on these debates came at a conference organized by the Lunar and Planetary Institute at Houston in February 1994. Two other major conferences, in 1981 and 1988, held amid the snowdrifts of Snowbird in Utah were the occasion for exciting encounters and acerbic debates over the Alvarez hypothesis. Published respectively in 1982 and 1990, the papers presented at these conferences fill two volumes that are indispensable for any student of the KT boundary. They clearly show that despite the efforts of a very small minority of dissenters, led by Chuck Officer and Dewey McLean, the impact theory has quickly become predominant by a wide margin, at least in the USA. Late in February 1994, the researchers would again come flocking; not to Snowbird this time, because that winter sports' resort had grown prohibitively expensive and the weather was bad to boot, but to the clammy plain of Texas.

Before the conference, Keller, Smit, and some of their associates had organized a field trip so we could examine the famous northeastern Mexican sections at and around Mimbral. And so several dozen scientists from an extremely varied assortment of disciplines and universities (Fig. 8.2) found themselves attentively listening to heated arguments between paleontologists, stratigraphers, and sedimentologists of the two camps, the one led by Gerta Keller and the other by Jan Smit. According to the latter group, the 7 m of this strange rock before our eyes had been laid down in less than a week, while according to the others, many thousands of years had passed in the process. The debate was summarized on the spot by the pithy

Figure 8.2
Field trip to Mimbral section, northeast Mexico, February 1994. Among those present, a number are mentioned in this book: Jan Smit, Bob Ginsburg, Virgil Sharpton, Hans Hansen, Bruce Bohor, Michael Rampino, Robert Rocchia, Philippe Claeys, Wolfgang Stinnesbeck, Al Fisher, the author, Gerta Keller, Bill Glenn . . . .

question, "100,000 seconds or 100,000 years?" My impression as a nonspecialist is that while none of those involved spoke without some authority, quite simply none of the disciplines in question was capable, given the available data, of distinguishing between these two durations: both of them equally brief in the light of our ability to measure time yet so very different in their dynamic consequences. One method might have provided an answer: paleomagnetism. Some Mexican researchers had, in fact, cited preliminary measurements in which they observed a reversal of the magnetic field between the bottom and top of the section. Now, a reversal cannot take place in less than several millennia.[12] If correct, this finding would rule out the tsunami hypothesis. So I took advantage of the visit to collect a few

12 Which makes it one of the fastest global geological events of internal origin.

specimens, which our student Yang Zhenyu measured at our laboratory in Paris. Superbly magnetized, with the reversed polarity and orientation of the magnetic field at the KT boundary typical for North America, the specimens showed no sign of a reversal, so the question remains open at the end of the 1990s.

The Houston meeting

During the subsequent conference in Houston, Bob Ginsburg was supposed to present the results of a series of "blind tests" intended to settle a number of the controversies that had arisen over the past few years. One of them addressed the extent of the distribution of iridium and shocked quartz in the Gubbio section (very limited according to Alvarez, more extensive according to Officer). Another concerned the duration of the extinctions in the El Kef section (long according to Keller, very short according to Smit). In each case, multiple specimens had been collected and distributed to different laboratories with no indication of their stratigraphic position. Bob Ginsburg would summarize the findings. Unfortunately for him most of all, but also for us, he fell down a flight of stairs on returning from the field and was unable to present his conclusions. Never mind! Gerta Keller and Jan Smit had the data, and Ginsburg sent his documents to his colleague Al Fisher, with a request to fill in for him on a moment's notice. And so we were able to read a poster presentation by Gerta Keller and witness an oral presentation by Jan Smit, each concluding that the observations incontestably showed they were right! As for Fisher's presentation, it was not very clear on any point but one: the extent of the anomalies at Gubbio was indeed quite limited. In what might have been considered an astonishing turn of events, when the two journals *Science* and *Nature* reported on this colloquium, they echoed only Smit's viewpoint, which admittedly was shared by the great majority of the audience. But in reality this was not so surprising; the journalist from *Science*, Richard Kerr, had been persuaded since 1980 of the correctness of the impact theory. As for *Nature*, which probably failed to notice this, the reporter the journal chose to write up the results was Smit himself.

The results of the blind test were finally published in January 1997 as 40 pages of the journal *Marine Micropaleontology*. In this very unusual exercise, two paleontologists concluded in favor of Keller and gradual extinction, and two in favor of Smit and catastrophic extinction at the iridium level. Bruce Masters from Fairbanks was adamant that the "overall extinction pattern does not fit that of a single catastrophic event," and even insisted that the massive extinction at the KT level might still be nothing more than an indication that part of the section was missing. Richard Olsson from Rutgers, however, was quite sure that "the pattern of occurrences of Cretaceous taxa across the KT does not support the gradual or stepwise extinction pattern." He even concluded that 12 out of 14 species Keller had reported as extinct below the KT actually reached the boundary.

For their part, in their summary analyses of the data provided by the four independent blind testers, both Keller and Smit agreed that part of the problem was the very limited coordination among the testers' approaches to the identification of the species themselves: for more than two-thirds of the species, the testers did not even use the same names! Gerta Keller concluded that patterns of species occurrences were in better agreement with her own stepwise scenario. Jan Smit, on his side, went on to group species into categories of synonyms, concluding that all species did reach the KT. Different testers, he pointed out, found different results only for rare species, for which it would be easy to miss the few individuals present in the small samples (an example of the Signor-Lipps effect).

So the blind test left the two main protagonists very much where they were before, believing that their early views had merely been further vindicated. This fascinating and all-too-rare exercise has clearly not resolved the controversy, but it does tell us more about paleontologists and paleontologic methods. As Bob Ginsburg concludes, there should have been far more samples than the six levels collected at El Kef, and the participants should have agreed on a common taxonomy prior to the exercise. It was this lack of consistency in taxonomy among the testers, and the rarity of many taxa, that made the test inconclusive. My own cautious conclusion (as a nonpaleontologist) is that the gradual aspect of the extinction prior to the KT at El Kef must be better documented and requires further confirmation, whereas the gradual aspect of the pattern after the KT seems better established. Clearly, there was a catastrophic event at the KT and major and lasting unrest afterwards.

At the Houston meeting, the supporters of the impact theory were so much in the majority that only they took part in the principal discussions. Also Camargo presented a review of the early evidence on the Chicxulub crater that was quite remarkable. For example, Hildebrand and Sharpton had a standoff over the question of the diameter of the crater of Chicxulub (Fig. 8.1): under 200 km according to one, almost 300 km according to the other. For my part, I remember from these debates two private conversations with Walter Alvarez.[13] After noting the latest summary I had drawn up of the ages of the traps and extinctions (see Chapter 6), he allowed that the correlation was so good that it would be difficult to dismiss volcanism altogether. He seemed prepared to accept that many extinctions could be linked to traps, though not the one at the KT boundary, for which the evidence of an impact was by now blatant. According to Walter Alvarez and Frank Asaro, the iridium anomaly at the KT boundary is a unique phenomenon in the entire Phanerozoic Era, i.e., the past 600 Ma. The exceptional size of the Chicxulub structure (if we believe Sharpton) can moreover be explained better by an even larger extraterrestrial object, such as might fall to Earth once every billion years (see Fig. 2.6, p. 35), than the smaller object, occurring every 100 Ma, that Luis Alvarez had cited.

At the same conference, the Indian cosmochemist Narendra Bhandari and his associates from Ahmedabad announced that they had discovered the iridium layer and the KT boundary between two well-dated flows of the Deccan Traps, in the district of Kutch. This fascinating observation cried out for confirmation by independent teams; it would demonstrate that the fall of the object responsible for the iridium was contemporaneous with the Deccan volcanism, which had begun several hundred thousand years beforehand. A team was quickly formed to make a detailed study of the Kutch sections. I recruited Robert Rocchia for iridium and the spinels, Gilbert Féraud and Corine Hofmann for the dating, and Yves Gallet for paleomagnetism. Bhandari accepted the idea of such a joint Franco-Indian

13 I had worked in California with Walter Alvarez years before and we had remained on excellent terms. Besides appreciating his many scientific contributions, I felt fortunate that our relationship always remained friendly regardless of where our scientific views led us. This seems to me to have been a pleasant exception in a climate in which intellectual arguments led for many to a personal quarrel.

project. The mission took place in April 1995, and we reported the first results at the March 1997 meeting of the European Union of Geosciences in Strasbourg.

We first found that the geology in Kutch was far from simple. There were few sections where the sequence of flows and intervening sediments could be mapped without ambiguity. Rocchia has not found any spinels but confirms a significant iridium anomaly. He considers this unmistakable proof that the impact signature is present, though somewhat altered by later perturbations in the sediment. Dating of the lava and paleomagnetic analysis show that the upper flows are very close to KT age and reversely magnetized. The lower flows are much more altered but appear to be half a million years older and normally magnetized. This is in full agreement with our earlier analysis of the magnetic stratigraphy of the Deccan Traps and does appear to confirm that the impact occurred while the eruption of the traps was in full swing!

9

Controversy and coincidence

Any attempt at reconstructing a process of scientific research undoubtedly tells as much about the researchers as about the object of their inquiry. People often speak of researchers "seeking the truth." Since Popper, we have realized that the ancient notion of truth should be replaced by that of a "falsifiable" model: the purpose of research is to construct a model consistent with the knowledge available at a given moment. Even if not accepted by all, this model will remain acceptable unless and until someone finds a key experiment or observation that contradicts it. It is in this sense that, for simplicity's sake, we retain the word truth instead of changing terms. Here it merely means a model that has not yet been proven false. But researchers are doing more than simply seeking this kind of truth. Filled with impulses of their own, bystanders and participants in a ceaselessly evolving story, they are obviously fallible. And less lofty human passions sometimes lurk behind the controversy.

Conflicts and bullyragging

In our quest for what caused the disappearance of the dinosaurs, and the mass extinctions in general, we have seen two schools of thought emerge, supporting two very different and *a priori* irreconcilable truths. Each of the actors has certainly had his or her own different motives in the pursuit of truth and the search for an answer. The thirst for knowledge has gone hand in hand at times with ambition. Less noble motivations have doubtless not always been absent, especially in those countries where the pressure to publish and be the first to achieve a significant (and "salable") result was the most intense.

No doubt because the number of researchers engaged with this problem was greatest there, and the competition there the most grueling, the sharpest debates took place in the USA. With his strong personality, Luis Alvarez was no stranger to these disputes. Thanks to the strength of his convictions and the quality of his reasoning and observations, the partisans of the asteroid theory soon won over the community of geophysicists, geochemists, and astrophysicists – those who were the most readily inclined to deal with mathematical, physical, or chemical concepts using quantitative models.

Geologists and paleontologists, who prefer to be closer to field observation, held back for a longer time.[1] When the impact theory was first formulated, the great ocean specialist Cesare Emiliani denied that a period of darkness and intense cold could have existed, on the grounds of the biological evidence: too many species had survived.[2] Bill Clemens, Walter Alvarez's neighbor in Berkeley, long contended that the extinctions at the end of the Mesozoic took more than 10 Ma. His explanation for them drew upon very modern ideas: he believed that the dynamics of ecosystems and the interdependence of species are governed by nonlinear laws. Therefore, small causes, small changes in the environment, might have produced large effects and major extinctions that to us appear abrupt. As we have seen earlier, these observations would attract acerbic buckshot from Luis Alvarez.

We might say that by the mid-1980s, the "impactists" had gained the upper hand with the professional conferences and funding agencies, to the point that some partisans went so far as to ridicule their adversaries and interfere with their careers. Several major scientists, particularly supporters of the volcanic hypothesis, suffered. For example, from 1985 on, McLean's work on volcanic emissions and the biological pump met with a profoundly unfair reception, which became an undeserved professional and psychological burden.

Another victim of these verbal scuffles was Chuck Officer. For almost ten years he had been a primary counterweight to Luis Alvarez, and it would be impossible to overstate the importance of his role throughout this scientific adventure. The articles he wrote with Chuck Drake in 1983 and 1985 made him one of the essential players, both stimulating and often very much to the point, in this controversy. Officer was not born yesterday. A geophysicist of world

1 With some exceptions, such as Stephen Jay Gould, author of a theory of evolution called "punctuated equilibrium" that gave him a ready understanding for catastrophic scenarios.

2 The idea that surviving species contribute at least as much information as those that disappear is central to paleontologic research today and shows great promise.

renown, he had contributed to oceanic exploration and the birth of plate tectonics and had made a fortune in petroleum research. And he was hard-nosed enough to hold his own against a scientific rival like Alvarez. Yet in March 1994, he bade a weary and bitter "farewell" to science journalists:[3]

Your having been interested in Cretaceous/Tertiary (K/T) affairs, I have enclosed for your perusal a manuscript which will be the basis for a presentation that I will make (. . .). The whole experience of the recent studies in the Caribbean has been bizarre – almost unbelievable (. . .). I find the whole thing embarrassing to geology much the same as the polywater controversy was an embarrassment to chemistry. The party is over. As these recent studies . . . sift into the scientific community, I hope that the whole K/T thing will become a no news item and scientists can shift back to doing science on the interesting subject of mass extinctions without the acrimony that the K/T debate has engendered. For myself, my work on the K/T is done with this manuscript. I have returned to and will continue to work on environmental science.

Knowing and telling

In science, an idea often surfaces, nobody quite knows how, on the basis of a few measurements or observations. But this is not enough. The idea has to be shaped, put through an initial battery of tests, written up, and finally published. It must become known and debated and make its way through the mill. The newer and more original the idea is, the more difficult this indispensable last phase will be. About the difficulty of getting the new quantum physics accepted, Max Planck once said, not without cynicism, that one does not convince the old guard of the validity of a new theory, one merely waits for them to die (and for enough of the younger generation to be trained in the new ideas). For a scientist, the standardized vehicle of thought is the written article, which will be published only after being

---

3 Letter dated March 4, 1994 to journalists at the *New York Times, Washington Post, Boston Globe*, Associated Press, *Science, Nature, New Scientist*, etc. Copy sent on the same day by Chuck Officer to the author. Chuck Officer finally summarized his views of the debate in C. Officer and J. Page, *The Great Dinosaur Extinction Controversy*, Reading, MA, Addison-Wesley, 1996.

tempered in the fire of criticism from a series of anonymous readers, chosen from among the author's peers. The final decision to publish is made by an editor-in-chief, who assumes his share of the responsibility along with the authors. A result announced orally at a conference does not have the same value and cannot normally be cited by a fellow researcher pursuing some other matter.

Unfortunately, given the proliferation of publications, researchers must try to impress and convince not only their own colleagues but the general public (the citizens whose taxes fund research in most countries) and the decision-makers (those who have a more direct influence on research funding yet no longer have the time to read everything and are just as sensitive as others to fashion and the *vox populi*). So more and more often scientists go ahead and release their results to nonspecialized journals first or even turn to television. This practice can only be deplored. In the long run it may threaten the integrity and the very value of scientific research itself. Let's be quite clear here: high-quality popularization of science is essential, but it must follow and not precede publication, which must observe professional standards. The controversy over the KT boundary has not escaped these excesses, particularly because the subject – involving dinosaurs as it does – is popular and commercially interesting.

The media has played an important and sometimes questionable role in choosing to publicize one theory over another. Not only editors of magazines for a broad public, but even those of some major scientific journals like *Science*, quickly opted in favor of the impact theory.

What is a catastrophe?

Why does the public apparently find an asteroid so much more glamorous than a volcanic eruption? The answer seems to me to lie in our anthropocentrism. An impact like the one the Alvarezes conceived is certainly a catastrophe on the geological time scale. But the idea of an impact feels just as catastrophic, and even almost unimaginable, when viewed on the scale of a human life. Its description, bandied about in the media, can spark the imagination, which is free to associate it with scenarios of a nuclear holocaust that appear in a great many films and books. Asteroid impact has been the central topic of

two competing films released in 1998. Moreover, the climatic consequences are very similar, as we have seen. Just about everyone
remembers the nuclear winter scenario constructed by atmospheric
scientists and planetologists in the early 1980s, but how many know
that this scenario was inspired by the impact winter, and thus directly
by the 1980 article in *Science*? In many people's minds there is an
unconscious and confused association between an atomic explosion
and the fall of a meteorite. So these ideas are easy fodder for journalists, who recognize a best-seller when they see one.

The same cannot be said of the volcanic scenario. Though
described several times in the major media, it doesn't sink in, it
doesn't "hook" people. And besides, people are used to big volcanic
eruptions, or at least have seen them enough on television and in the
movies that the thrill has worn off. Spectacular and unsettling, of
course, but a "small" catastrophe after all, one that cannot compete
with a real explosion, whether extraterrestrial or of human design.
Our perceptions are distorted by a double inequality (in the mathematical sense). The duration of an impact can be measured in seconds. So it is "infinitely" shorter than a human life (by nine orders
of magnitude, or a factor of one billion). By contrast, the Deccan
eruptions, which lasted hundreds or even thousands of centuries,
seem "infinitely" longer (by four or five orders of magnitude in the
opposite direction, or a factor of 10,000 to 100,000). To the non-
geologist, a meteorite impact is an almost instantaneous catastrophe
of extreme intensity. By comparison, unless one lives near an active
volcanic zone, an eruption, even a powerful and protracted one,
seems rather a minor event. And yet one must realize that in terms
of the long scale of geological time – the scale that governs the Earth's
existence and the life spans of species, and that the reader must be
getting used to by now – the eruptions of the traps were indeed
catastrophes, hard to imagine because none has occurred within
human memory. The last of them, the small Columbia Traps, in
North America, ended 16 or 17 Ma ago and seem to have had no
global effect on the biosphere. The preceding eruption was that of
the Ethiopian Traps, which as we have seen marked the end of the
Lower Oligocene, 30 Ma ago. The last truly murderous eruption for
a very great many species was that of the Deccan, 65 Ma ago. It takes
the geologist's perspective, reinforced by a certain feeling for long

periods of time,[4] to realize that both of the scenarios I have been developing throughout this book do indeed deserve to be called planetary catastrophes.

Death from the sky

What is my position today, amid this exciting debate that has been unfolding since the late 1970s? You can easily tell from the observations I have tried to summarize and the new developments I have related above.

I now think that quite an exceptional extraterrestrial object must indeed have struck the Earth 65 Ma ago. As I said above, after accepting this hypothesis for several years as a dilettante, I rejected it in later years on the grounds of intellectual parsimony, or what some call "Occam's razor."[5] It did not seem reasonable that two exceptional events could have coincided in time. The original "proof" of the impact, on closer study, seemed insufficiently definitive to me (prior to the discovery of the Chicxulub crater and Doukhan's latest work on shocked quartz). The Deccan volcanism, by contrast, was quite real. We had just established its extraordinary intensity and, what's more, it was the right age.

Several years have passed. The iridium anomaly and shocked quartz remain unique to the KT boundary. For now, at least, they cannot be explained satisfactorily by the supporters of volcanism. So at the end of the first quarter, the score is *impact, one.*

But though the other boundaries of geological ages have generally not yet received the same attention as the KT boundary, none of them seems to include abnormal quantities of iridium that cannot be explained by purely terrestrial mechanisms of production and concentration. Geochemists have recently discovered an iridium anomaly, which seems to be global, at the boundary between the Devonian and the Carboniferous Periods: 360 Ma old, this extinction is one of the five great ones of the Phanerozoic and precedes the extinction at the end of the Permian. But given the absence of shocked quartz and

4 Well beyond the long time-constants of human history laid out by the French historian Fernand Braudel, the full consequences of which are already so difficult to imagine.

5 The principle that given two models equally capable of accounting for observations we should prefer the one that is most economical in terms of hypotheses or parameters.

microtektites, and the presence of very reduced (oxygen-depleted) black shales, we must conclude that this anomaly could have been caused by purely terrestrial chemical phenomena that occurred at the time of deposition or during diagenesis of the sediments. So no undisputed impact at any of the boundaries other than KT.

Death from the mantle

The traps are extraordinary geological objects whose importance has long been underestimated. They supplement the relatively calm picture painted by "normal" plate tectonics. These tectonics are not sufficient to rid the Earth of all its heat. Episodically and irregularly, every 20 or 30 Ma or so, an immense bubble of material from the mantle becomes unstable, rises to the surface, and bursts. Its emergence triggers gigantic explosive eruptions and finally lays down millions of cubic kilometers of basalt within a few tens or hundreds of thousands of years. Some ten traps have been identified from the past 300 Ma. Seven of them (see Chapter 6) coincide with seven mass extinctions, and in particular with the two largest ones. The more refined the dating, the better the correlation looks. This volcanism can account qualitatively, and in some cases quantitatively (very roughly speaking), for the many anomalies found in the stratigraphic series that have preserved a fragment of the record of these events. Eruptions of such magnitude seem able to affect the biosphere by injecting ash, aerosols, and gases, and they can probably cause darkness, temperature variations, and acid rain. It is clear that rarely, but sharply, the internal dynamics of the globe affect the evolution of species. The score at half time: *traps, seven.*

Seven traps and one impact

Marine regressions, able to lower sea level by 200 m within a few hundred thousand or a few million years, also coincide with many boundaries whose traces they have often nearly erased. Though the mechanism has yet to be elucidated, it seems easier to link these regressions to convection in the mantle and to plate dynamics than to possible impacts that they would have systematically, and miraculously, to precede by about a million years. However, plate tectonics (a "normal" mode of convection) and the birth of hot spots (a more "exceptional" mode of convection) may well have a shared first cause, internal to the mantle, but different response times. A slowdown in sea-floor spreading, leading to a generalized drop in sea level (as the mid-ocean ridges would take up less space), might precede the formation of the instabilities in the deep mantle, that lead to the eruption of traps. The crustal uplift caused by the thermal effects of the head of the plume likewise would entail a (regional) retreat of the seas. So it is not absurd to imagine a shared mechanism, or at least a shared first cause, for marine regressions and the traps. It may also be that our stratigraphic record of sea level fluctuations has to some extent been obscured by later alteration. Part of the drop in sea level might have been faster and simply be linked to major phases of ice-cap growth or formation, in association with a trap-generated volcanic winter. This appears to be the case for the most recent traps, where the record may be more easily decipherable: the Ethiopian traps, 30 Ma old.

Moreover, few of the supporters of the extraterrestrial hypothesis insist any longer on an impact origin for boundaries other than the KT boundary. Must we allow that on a single occasion, while the Deccan Traps were already erupting and the biosphere was sorely put to the test, an impact occurred at the same time, dealing a further blow to species so severely tried already? This is what seems most likely today. The proof may well be provided by the discovery of the iridium-rich layer sandwiched between two flows from the Deccan Traps, in the Kutch province of India (see Chapter 8). But asteroids no doubt played an infinitely more important role during the Earth's first billion years. If we believe Jay Melosh, some ten major impacts of objects as large as one-tenth the volume of the modern Earth then caused our planet to change catastrophically, each impact either tearing off and melting a major piece of the already formed object or adhering to it and enlarging its mass. The Moon apparently resulted from one of these shocks. No form of life was possible then. Undoubtedly it was only after the end of this gigantic bombardment, perhaps 4 billion years ago, that Life could arise. A good deal of debris was still left between the planets, and very large impacts must have continued to play a major role for some time. But surely not in the past billion years.

The final score (or rather the current score, since we may perhaps only be at the end of the third quarter): seven to one.[6] The catastrophic eruptions of the traps do seem to have been the principal agent that episodically turned the evolution of species in a new direction, unexpected until then but mandatory thereafter. Only once in 300 Ma, at the KT boundary, did species that were already hard pressed suffer the additional catastrophe of an impact by an asteroid or comet. The respective repercussions of these two events for the climate and biosphere have yet to be worked out in detail. And let us recall that among the clearly dated craters, none seems to have had any visible effect on the diversity of species except Chicxulub.

6 In Chapter 6, we discussed recent evidence of another impact near the Jurassic–Cretaceous boundary. Much work is still needed to confirm this.

10

Improbable catastrophes and the flukes of evolution

A number of works have explored the climatic consequences of the impact; somewhat fewer have discussed the eruptions of the lava floods of the traps, those "staircases of fire."[1] We have had a glimpse of these in the preceding chapters. Each of the catastrophic scenarios predicts rather similar climatic events: dust, acid rain, and cooling, followed over the longer term by global warming. Darkness, noxious gases, forest fires, survivors that take shelter by creeping into the smallest of burrows – all this is certainly food for the imagination. As we have seen, the volcanic and impact catastrophes differ only – and this is not unimportant – in their duration: several millennia in one case and less than a year in the other. The fact that the two perturbations are so similar makes it even more difficult to try to distinguish between them solely on the basis of the physical and chemical record preserved in sediments.

The quantities of gas poured out by volcanism into the atmosphere at the end of the Cretaceous Period must have been considerable: possibly ten million million metric tons of sulfur dioxide, just as much carbon dioxide, and 100,000 million tons of halogen gases, particularly in the form of hydrogen chloride. The duration of the crisis would determine the rate at which these gases were injected into the air. This injection rate is a very important parameter. Humans are injecting the same potentially toxic gases into the atmosphere, at rates already in excess of the natural processes that are active during the rather calm period we live in. These rates are indeed comparable to those that may have been produced by an impact or the laying down of a giant lava flood.

1 See Chapter 3, Note 11.

Climatic catastrophes: is the past the key to the future?

And thus geologists' discoveries link up with extremely current concerns. For the first time in the history of the Earth, a living species is able to produce quantities of products – solid, liquid, or gas – on the same order of magnitude as those produced naturally by the entire Earth itself. But the durations and output rates are not the same. For amounts that may have taken the Earth hundreds of thousands, or even millions, of years to produce, humans have taken only 100 years, and in many sectors production continues to grow exponentially.

This is not the place to construct a predictive model of the climate based on the current production of aerosols and gases – and at any rate, I would not be able to do so. Moreover, the uncertainties are still considerable; even the famous greenhouse effect is still a matter of debate. Some scientists say there is no doubt that the increase in the concentration of carbon dioxide in the lower atmosphere, caused by industrial activity and artificial heating, is responsible for a warming of our planet that has still barely begun. But others, such as Yves Lenoir and Claude Allègre, cite the good correlation between the ancient carbon dioxide contents measured in cores drilled from Arctic and Antarctic ice[2] and the temperature over the last few glacial and interglacial periods. These two distributions are themselves correlated with the Milankovic cycles.[3] Since these cycles relate to purely mechanical variations – astronomical variations in the different parameters of the Earth's orbit and the inclination of its axis of rotation relative to the Sun and to the plane of the ecliptic – these scientists hold the causal relation to be exactly the reverse of the greenhouse effect: astronomical parameters govern the amount of sunlight that reaches the Earth and, thus, the mean temperature, which in turn regulates the concentration of carbon dioxide, by simple dissolution into or outgassing from the ocean.

So it may not be significant that the mean flux of carbon dioxide produced by the Deccan Traps is on the same order as current human production. But the instantaneous flow rates may have

2 See for example C. Lorius, *Glaces de l'Antarctique: une mémoire, des passions*, Paris, Odile Jacob, 1991; Y. Lenoir, *La vérité sur l'effet de serre*, Paris, La Découverte, 1992.

3 See Chapter 1, Note 12.

been much higher, and we have seen what consequences can be deduced from this fact, particularly the tenfold multiplication of the carbon dioxide content in the event of the death of the biological pump. The (quantitative) effect is even less well known for sulfur and the halogens, but the orders of magnitude suggest considerable climatic repercussions: acid rain and destruction of the ozone layer. So geologists can currently offer climatologists the boundary conditions, and the long-term historical perspective, that the latter lack. For a more reliable estimate of emission rates in the great catastrophes that led to mass disappearances of species, we need a very high-precision method of measuring time, one more refined than we can currently achieve. A one-second impact will not have exactly the same consequences as a year-long eruption. And since volcanism is by nature an episodic phenomenon, the number and spacing of eruptions over time is an essential factor to determine, because these may result in interacting phenomena, a saturation, or even the onset of a nonlinear regime.

Without meaning to sound pessimistic, I believe the ancient catastrophes whose traces geologists are now exhuming are worthy of our attention, not just for the sake of our culture or our understanding of the zigzaggy path that led to the emergence of our own species, but quite practically to understand how to keep from becoming extinct ourselves. That said, virtually every species has eventually died out, generally after a few million years at the very most.

The third crisis

Some authors think we have already entered a new period of mass extinction.[4] Around 6 Ma ago, ice little by little invaded the Antarctic and then gradually extended into even vaster regions. Glaciation became sufficiently widespread 2 Ma ago to be characterized as a new geological stage, the Pleistocene. The evolution of the human species is profoundly linked with the variations in climate during this period. The Pleistocene saw the extinction of two-thirds of all mollusks, gastropods, and bivalves in the Western Atlantic and Caribbean. The

4 See Peter Ward's fervently argued book *The End of Evolution*, New York, Bantam Books, 1994.

recession of sea level, associated with the growth of the glaciers, was surely responsible in part. (These variations may not be related to those observed at the end of the Paleozoic and Mesozoic, which as indicated above may have been linked to variations in the rate of sea-floor spreading.) Mammals in Africa were likewise severely affected, as were one-third of the mammals in North America. Far more recently, 11,000 years ago, two-thirds of the large mammals surviving on the American continent (North and South) disappeared, quite suddenly. Many researchers see this as the consequence of the arrival of humans by way of the Bering Strait.

Far more recently again, human activity has been the cause of many extinctions. These are particularly visible on Pacific islands like Hawaii, where the arrival of our species a little less than 2000 years ago devastated the original flora and fauna. The arrival of Captain Cook and Western civilization, beginning in 1778, unleashed new waves of extinction. The same tale can probably be told of Madagascar, New Zealand, and many more places.

Until recently, the total number of species currently populating the Earth was estimated at 5 million. With the discovery of the richness (and the smallness) of some habitats in coral reefs and tropical rain forests, the estimate has by now reached something like 50 million! Simply by cutting down the rain forests, we are probably wiping out some species even before we have a chance to discover they exist. In all, 100 species are thought to become extinct every day.

In *The End of Evolution* (see Note 4), Peter Ward does not hesitate to suggest that the Pleistocene represents the third great mass extinction in the Earth's history, after those that marked the end of the Paleozoic and Mesozoic. But let us note that, however striking it may be, the title of his book is misleading. The mass extinctions mark, not the end of evolution, but its major changes of direction. The reality of this "third crisis" is still, as you may imagine, open to debate. Its extent and duration are disputed. So it is difficult to compare the extinction rate I mentioned above with the extinctions revealed in the fossil record. However imperfect, our knowledge of living species is far superior to our knowledge of fossil species. More than 70 percent of all species leave no fossils behind, and even among those that can be preserved, the vast majority of individuals vanish forever without a trace. We do not have, so to speak, a single fossil clue to animal species analogous to those species that are being discovered daily in the tropical jungle.

If this third crisis is real, what are its causes? It is hard to untangle the role of climate from that of humans. Two schools have squared off against one another, each supporting the idea that one of these agents is the primary cause. Ward notes that Lyell himself could already write in the mid-nineteenth century:[5] "We must at once be convinced, that the annihilation of a multitude of species has already been effected, and will continue to go on hereafter, in a still more rapid ratio, as the colonies of highly civilized nations spread themselves over unoccupied lands." Using amino acids in the egg shells of the gigantic Australian bird *Genyornis*, G. H. Miller and colleagues[6] have shown that these birds disappeared rather suddenly 50,000 years ago, at a time of no particular climate change but very close to the time of arrival of humans in Australia. Human population has evolved with a rapidity and intensity unparalleled in the history of evolution. So the crisis began well before the industrial era, so often blamed as the sole cause; this underscores the importance of a detailed study of the great crises of the past if we are to understand future alterations in our environment. In a world already weakened by eruptions, glaciation, the presence of humans, and so forth, what would be the consequences of an additional abrupt event like an impact, a sudden eruption, the wastes from such a brief time as the industrial age? Could such an event tip the system over into mass extinction?

Improbable catastrophes

Several scenarios, as we have seen, continue to be cited in the attempt to account for the upsets the blue planet suffered at the end of the Mesozoic. Some say nothing really catastrophic happened at all. On a scale of some 10 Ma, the seas withdrew, then advanced again, and the changes that this ebb and flow provoked in the extent of emerged land, the ocean currents and the climate are enough in themselves to account for the extinction of old species and the appearance of new ones.

5 Cited by P. Ward (see Note 4), p. 197.   6 G. H. Miller *et al.*, *Science* 283, 205, 1999.

Set off against this "uniformitarian" theory – supported by only a very small minority today[7] – are two opposing "catastrophist" theories that we have discussed here at length. One theory invokes a truly dramatic catastrophe, a cataclysm even when viewed on the scale of a human lifetime: that of an asteroid or comet impact. The murder weapon is a large extraterrestrial body. So the first cause of the extinctions must be sought outside the Earth. It has no simple and direct, or at any rate causal, relation with geology. No wonder a good many geologists initially viewed this as an unacceptable appeal to a *deus ex machina*. Yet planetary exploration, and the new perspective of the Earth it has given us, shows beyond a shadow of a doubt that impacts are geological agents that we can no longer ignore, although their role was doubtless far more decisive early in the Earth's history than it has been more recently.

The second catastrophist theory, volcanism, invokes colossal eruptions of continental basalts. The partisans of the asteroid have naturally tended to downplay the importance of this factor. Yet it is now clear that the traps represent the greatest episodes of volcanism on Earth, and that they were almost a hundred times briefer, and their intensity, therefore, a hundred times greater, than was thought just a few years ago. These two types of catastrophe certainly existed, and the two schools, however irreconcilable they may have seemed at times, are undoubtedly both on a right track. They carry us to the farthest outposts of our scientific disciplines, physical frontiers to which human beings can hope to travel only in thought: the depths of space and the bowels of the Earth. Ever since Kühn, we can understand how natural it was that many researchers, as with any major scientific advance, would try to resist such novelties.

The passionate debates that have raged around this problem in the geosciences since the late 1970s are a new reincarnation of the centuries-old debate between catastrophists and uniformitarians.[8] Catastrophism, associated with the name of Cuvier, is a system, a globalizing theory. But unlike uniformitarianism, it affords us no

---

7 Two remarks: a theory is not false merely because it is supported by a minority. And the seas unquestionably evolved, and there were consequences. But in my opinion we cannot deny that there was another cause, much briefer in nature.

8 See Chapter 1, and also Anthony Hallam, *Great Geological Controversies*, Oxford, Oxford University Press, 1983.

method, no research technique. Indeed, Cuvier himself applied the uniformitarian method to show that the Tertiary layers of the Paris Basin represent cyclic alternations of saltwater and freshwater conditions. But where more recent times were concerned, he invoked supernatural cataclysms and deluges, assuming that no agent of nature today could account for what he observed.

A good many geologists of the nineteenth century in fact recognized the need to apply uniformitarianism conjointly with (a pinch of) catastrophism. In their excellent book on the evolution of the Earth, Dott and Batten[9] wonder how the controversy could have lasted so long, when Playfair could write as long ago as 1802 that:

Amid all the revolutions of the globe the economy of Nature has been uniform, and her laws are the only things that have resisted the general movement. The rivers and the rocks, the seas, and the continents have been changed in all their parts; but the *laws which describe these changes,* and the rules to which they are subject, have remained invariably the same.[10]

In the 1830s, Whewell expressed the same views in the clearest terms, and to my way of thinking the most modern too, when he declared that natural laws and geological processes are certainly universal in their physical and chemical aspects, but nothing can guarantee that they will apply at a constant rate or a steady speed.[11]

In order to enable ourselves to represent geological causes as operating with uniform energy through all time, we must measure our time by long cycles, in which response and violence alternate; how long must we extend this cycle of change, the repetition of which we express by the word uniformity? And why must we suppose that all our experience, geological as well as historical, includes more than one such cycle? Why must we insist upon it, that man has been long enough an observer to obtain the average of forces which are changing through immeasurable time?

Would it not be arrogant to imagine that our history, which represents only one ten-thousandth of the history of Life, and one-millionth of the history of the Earth, allows Man to sample and preserve the

9 R. Dott and R. Batten, *Evolution of the Earth,* New York, McGraw-Hill, 1981.
10 Quoted in Dott and Batten (Note 8), my italics.
11 Quoted by Anthony Hallam in *Great Geological Controversies,* Oxford University Press, 1983, p. 53.

memory of the full variability of the phenomena that may occur on the planet?

It was Lyell – pushing to an extreme the uniformitarian concepts he had introduced, no doubt in the effort to impose them by dealing a knock-out blow to the opponent – who practically denied the existence of an *arrow of time* in geology. He did not see the presence of History in any real sense. He believed the impression of change given by fossils is illusory, the result of a lack of preservation of higher organisms. He ridiculed catastrophists and belittled or ignored their essential contributions: stratigraphy, the direction of time, evolution. Many of Lyell's supporters did not share his extremism and on the contrary adopted the notion of a regular progression. In 1869, Huxley contended that no geologist of his day would stand up for the "absolute" version of uniformitarianism. According to him, the idea of unlimited geological time contradicted the second law of thermodynamics and the very notion of history itself.

Neither impacts nor catastrophic volcanic eruptions have any place in Lyell's vision. But, as Eric Buffetaut says,[12] the controversy between Cuvier and Lamarck does not merely set up an opposition between a retrograde belief in fixity and a growing evolutionism. Even if Cuvier, besides his attractive style, had according to Buffetaut a stricter method than his opponents, still as a general rule it would be the uniformitarians who would develop the tools and practices of modern geology over the course of the nineteenth century. Lyell was right in founding his methods on the principle of real causes and refusing to invoke supernatural forces. His rigorous and even extremist uniformitarianism was perhaps a tactical necessity to counter the catastrophists. Nevertheless, Cuvier was not wrong to emphasize the reality and importance of certain catastrophes. But no doubt it would have been disastrous to realize too soon that he was right. If the reality of sudden events with no modern equivalent had been accepted from the end of the eighteenth century on, many thinkers would have appealed to great unverifiable phenomena from the outset. A discovery has to arrive at the right time.

---

12 See Chapter 1, Note 6.

The broken line, or the drastically pruned bush

In a recent article celebrating the 21st anniversary of the theory of punctuated equilibrium,[13] the theory's primary authors, Gould and Eldredge, remark that in the natural sciences all great theories are based on the frequent repetition of the same kind of observation but do not necessarily imply the exclusivity of that observation. The gradualism of evolution is well documented for some species, as is the theory of punctuated equilibrium for others. The sticking point is to know which of the two is common enough to impose its rhythm and signal on the history of Life. Gould and Eldredge obviously make no secret of the fact that they believe observation supports a preference for their theory. I cannot resist drawing a parallel with our two catastrophist theories of mass extinctions: the "slow" theory of the mantle plumes and the "fast" theory of the asteroids. The two are not exclusive. And if we can speak of frequency where such small numbers are involved, the reader knows by now where my own sympathies lie, or more exactly which theory I think is more important to an understanding of the great changes in the direction of Life on Earth, on the grounds of "seven to one."[14]

As Gould and Eldredge say,[15] "contemporary science has massively substituted notions of indeterminacy, historical contingency, chaos and punctuation for previous convictions about gradual, progressive, predictable determinism." Mass extinctions have, in fact, often stricken down species that were extremely well adapted to their environment – as long as that environment remained within

13 This theory explains the absence of intermediate fossil forms between two distinct species (one descending from the other) not as the result of the poor quality and incomplete nature of the sedimentary record but by the fact that evolution takes place in very rapid phases separated by longer "stases," during which the ancestral form survives almost unmodified, with no gradual transformation. So the evolutionary tree looks more like a drastically pruned and espaliered bush. See for example Stephen Jay Gould, *Wonderful Life*, Norton, New York, 1990.
14 Or, more accurately, seven (traps) and one (impact).
15 S. J. Gould and N. Eldredge, *Nature*, 366, 223–227, 1993. See also Note 13 and Gould's,

book. In the preface of his ode to contingency, Gould explains his choice of the title *Wonderful Life*, an allusion to Frank Capra's film *It's a Wonderful Life*. The main character (Jimmy Stewart) has a guardian angel who replays life for him as it would have been if he had never existed and shows him the considerable changes in history that would have resulted from such an apparently insignificant event. For French *cinéphiles*, another good example of historical contingency can be found in the pair of films *Smoking and No Smoking*, by Alain Resnais, in which an act as minor as deciding whether to smoke a cigarette leads to a cascade of totally different events.

"normal and reasonable" limits. The media's popularized representation of worn-out, ill-adapted, obsolete, stupid, oversized, sluggish, gluttonous, fragile-egged dinosaurs has no basis in fact. Many of these great saurians were indeed the masters of their world, and some of their species were equipped to withstand almost anything except the sky falling on their heads. Being wiped out under such conditions (whether geophysical or astrophysical) can be chalked up to bad luck. But once their disappearance was complete, evolution resumed its "experiments" more vigorously than ever, and the consequences, unpredictable at the outset, became inevitable.

The eruptions of the traps, which were at least in part responsible for the principal extinctions, have punctuated the history of the evolution of species (with assistance from the occasional asteroid) and thus provide an excellent illustration of the contingent model Gould describes for the astonishing Cambrian fauna of the Burgess Shale, more than 500 Ma old (see Note 13). In these shales from the Canadian Rockies, we find delicate imprints from the soft parts (which are very seldom preserved) of marine animals, indicating the existence in the past of anatomical plans very different from those of the organisms that have populated the Earth since then and down to our own time. Whereas at least 25 different anatomical plans rapidly developed during the Cambrian Period, only four, none of whose future success might have been predicted, actually had descendants. One survivor, *Pikaia*, a chordate originally thought to have been a worm,[16] was perhaps the ancestor of all vertebrates. Nothing indicates that all those forms that died out, victims of a catastrophe of which no trace has yet been found, were any less well adapted to the Cambrian world than the few survivors. Like a broken line, "Darwinian" evolution thus seems punctuated, over the very long term, by catastrophes that wiped out some experiments that had nevertheless been highly successful, and opened the way for others. This recalls the mechanism of punctuated equilibrium proper,[17] which operates on a shorter time scale.

16 Its true identity was discovered by Simon Conway Morris and Harry Whittington. The chordates are organisms whose embryos have a nerve "cord," the starting point for a spinal column.

17 In a kind of nested system, operating over vastly different time scales.

Together with others, including Walter Alvarez, I have suggested that we should abandon the conventional expression "survival of the fittest," at least over this long time scale and for these major periods in which the entire face of Life on Earth is transformed, and instead use what seems to me the more appropriate term, "survival of the luckiest." Survivors are self-evidently fitter in crises, but these crises have nothing to do with the long-term conditions under which they had previously evolved. One cannot speak of adaptation to enormously rare events.

System Earth

The ferment of ideas that I have described in this book and that has kept the Earth sciences stirred up since 1980, is a sign of good health. It has given rise to a swarm of articles (the 2000 mark has been passed long since), to the point that it is now practically impossible to keep a complete reference list updated. No matter whether the aim was to defend or refute them, these ideas have prompted the acquisition of new data and the development of more and more precise analytical techniques. Some ideas may disappear in time. But high-quality observations and measurements will remain an irreplaceable asset, and any new theory will have to take them into account. This holds as true for the abnormal concentrations of iridium, or of shocked quartz and zircon, as it does for the measurement of the age of the traps and the discovery of their eruptive violence and geodynamic importance. Must we adhere to Darwin's substantially optimistic viewpoint? He writes: "False facts are highly injurious to the progress of science, for they often endure long; but false views, if supported by some evidence, do little harm, for every one takes a salutary pleasure in proving their falseness: and when this is done, one path towards error is closed and the road to truth is often at the same time opened."[18]

Since 1980, the Alvarezes' article has had the further happy consequence of spurring a true exercise in multidisciplinary cooperation. Understanding the physical, chemical, and biological events that

---

18 C. Darwin, *The Descent of Man*, 1871, Chapter 21.

happen in great mass extinctions requires contributions from a very large number of disciplines. All branches of geology (including sedimentology, stratigraphy, paleontology, and mineralogy), of geochemistry and geochronology, of geophysics, but also of biochemistry, organic chemistry, nuclear physics, astrophysics, materials science, fluid and shock mechanics, many branches of applied mathematics and computer science, and oceanic and atmospheric sciences have made their contributions, and this list is certainly not complete! We can no longer remain unaware of the language of these disciplines,[19] and the contributions they can make; yet at the same time it is impossible for any one person to master all of them. An uncomfortable position, this: being condemned to reconcile extreme specialization (in one's own discipline) with a superficial but obligatory knowledge of most of the others. The present book is an example. As a specialist in paleomagnetism, I have ventured out onto shifting ground, where others can doubtless easily catch me out in mistakes. But we have to take this step or abandon all hope of multidisciplinary understanding.

In our geosciences, as elsewhere, compartmentalization is harmful. Each individual must, through the medium of his or her own specialty, be able to contribute new, reliable, and useful data to other fields. In a way, one thus acquires the right to discuss proposed models with one's colleagues, often from other disciplines, and in brief to be a player in that exciting game called research. What luck, and what a unique adventure for a young student, to live through a scientific revolution, large or small – a questioning of knowledge that was thought cut and dried. My own generation has experienced the revolution of plate tectonics, the modern version of continental drift. But a great many very big problems remain unsolved, and the newcomer must not fear that his discipline will become extinct during this less thrilling phase while a new paradigm solidifies. Hot spots, impacts, magnetic-field reversals, and sudden isolated events have troubled what was becoming the placid surface of geodynamics. So

19 We often speak of the hard and the soft sciences, the latter being the human and social sciences in particular. Some years ago, when I was in charge of the funding, I suggested replacing this pair with the "inhuman" versus soft sciences.

Speaking of geology, David Raup wittily writes that it is more fun to do research in one of the "difficult" sciences than in the sciences commonly called hard.

many unstable, intermittent phenomena come together in this new way of looking at things, which elsewhere bears the name "deterministic chaos." "System Earth" is revealing itself to us in its profound unity. The Earth's rotation, the turbulence in its liquid core, disruptions in its mantle, volcanism and the climate, and finally the life of species, all these dynamic manifestations may be subtly linked. Over what time scale do these links have the most decisive effects? To what variations is the climate most sensitive? The tens to hundreds of millennia of the orbital Milankovic cycles? The centuries or decades of volcanic eruptions and human activity? The seconds of an impact? In the long history of evolution that has led to the world as we know it, the role of chance seems just as definitive as that of necessity. No doubt the party will go on. And many secrets remain to be brought to light.

Glossary

The footnotes also contain explanatory comments or sources of more details. Some items in the Glossary are adapted from R. L. Bates and J. A. Jackson (eds.), *Glossary of Geology*, 3rd edn, Alexandria, VA, American Geological Institute, 1990; *Oxford Illustrated Dictionary*, Oxford University Press; and *Dictionary of the Earth Sciences*, New York, McGraw-Hill.

29R The 29th major interval (also called magnetic zone or chron) with reversed polarity in the time scale of geomagnetic field reversals (very short events are neglected). The Cretaceous–Tertiary (KT) extinction and boundary occur during chron 29R.

Abatomphalus mayaroensis The latin name of a species of planktonic Foraminifera that characterizes the very last biozone of the Maastrichtian (and, therefore, of the Cretaceous *and* the Mesozoic) in many marine sediments. The *Abatomphalus mayaroensis* biozone lasted about 1 million years.

Abyssal plain The flat region of the deep ocean floor (approximate depth 4000 m).

Adiabatic change Characterizes a physical transformation in which material undergoes changes in temperature and pressure (and volume) without exchanging heat with its immediate environment.

Aerosol A gas that contains colloidal particles or droplets (solid or liquid, e.g. mist or fog).

Alteration A change in chemical or mineralogical composition owing to exposure to pressure, temperature, and chemical conditions other than those at which the (thermodynamically stable) material was formed. For instance, fluids passing through a rock will commonly alter certain types of silicate minerals (such as feldspars) into clay. Often synonymous with "weathering".

Ammonite A marine animal (cephalopod mollusk) with a coiled spiral shell that contains several compartments (or chambers) used to control swimming depth. The sutures marking the edges of chambers developed lobes and saddles with a characteristic, sometimes very complex and ornate appearance. They diversified in the Paleozoic and Mesozoic,

flourished in the Jurassic and Cretaceous Periods, and became extinct at the end of the Mesozoic. The *Nautilus* species is a presently living relative, with a very simple suture.

Andesite A volcanic rock composed essentially of a plagioclase feldspar called andesine together with more mafic constituents (pyroxene, hornblende, and biotite). Andesites are typical lavas from subduction zone-related volcanism and take their name from the Andes, part of the circum-Pacific fire belt.

Angiosperm Flowering plants (with visible reproducing system). They date since at least the Early Cretaceous.

Anoxic events Events recorded by sediments in which the ocean was largely depleted in oxygen, leading to a reducing environment with unusual chemistry and biota. Black shales and ratios of stable isotopes (for instance) provide evidence for these events.

Antimony (Sb) A (poisonous) chemical element that occurs in the Earth's crust and is relatively mobile in the near-surface environment.

Apennines A mountain range forming part of the axial spine of Italy, resulting from continental collisions during the Alpine orogeny in the Cenozoic.

Argon (Ar) The third most abundant element in Earth's atmosphere (at 0.9%; N_2, 78%; O_2, 21%). One of its isotopes (^{40}Ar) is a daughter product of radioactive potassium (^{40}K). This allows geochemists to date potassium-bearing minerals using the potassium–argon method.

Arsenic (As) A (poisonous) chemical element that occurs in the Earth's crust and is relatively mobile in the near-surface environment. It is also found in meteorites.

Arthropods Invertebrate animals characterized by a segmented body and protected by an external shell (or carapace).

Astatic magnetometer An early magnetometer (used in the 1940s to 1960s) that works on the principle of the attraction/repulsion of a magnet to a magnetic rock sample. The torque between the two is a function of the magnetic strength of the rock sample measured in a given direction.

Atmospheric plume A plume of upward moving warm convecting gases, aerosols and sometimes ash formed during the eruption of certain types of explosive volcanoes.

Basalt A dark-colored, sometimes olivine-bearing, volcanic rock, composed primarily of plagioclase and pyroxene silicates and produced by melting mantle material. It is the most abundant rock type in oceanic crust.

Basin and range A type of landscape that is created where the continental crust has been extended to form a series of linear ranges and basins, bordered by normal faults. A classic example of this is in the southwestern USA and northern Mexico where it is called the Basin and Range Province.

Bentonite A clay formed from the decomposition of volcanic ash.

Biomass Total mass of living organisms.

Bioturbation Disturbance of soft sediments by organisms that live and burrow in them.

Bivalves An animal, like a clam, that has two valves that open and shut (part of the mollusks).

Boundary layer A layer close to the boundary between two media with different physical or chemical properties. The term is used when a transport property is being studied (flux of heat or material . . .). For example, when internal heat passes from the Earth's core to the lower mantle, or from the upper mantle to the atmosphere and oceans, a boundary layer forms in which heat is dominantly transferred by conduction rather than convection (the boundary layers in these cases are respectively called D″ and lithosphere).

Brachiopods A phylum of marine, shelled animals with two unequal valves that filter seawater through tentacle-like arms. They thrived in the Paleozoic.

Brazil twins A crystallographic twinning pattern that occurs in quartz.

Breccia A rock with components that are angular coarse fragments formed by crushing along fault zones, or accumulation at the bottom of a talus following gravity sliding, or accumulation of eruptive material. These should be distinguished from conglomerates, where fragments are waterworn and rounded.

Bryozoan An invertebrate that builds varied calcerous structures and often grows in colonies.

Calcium carbonate ($CaCO_3$) Common in seawater and the principal constituent in limestone, which precipitates in the ocean. It forms the skeleton (hard parts) of many marine organisms. The most common mineral composed of $CaCO_3$ is called calcite.

Carbon cycle The distribution and exchange of carbon from reservoirs in the biosphere, atmosphere, and hydrosphere.

Carbon monoxide Produced during combustion, a poisonous gas.

Carbonate In geology, carbonate is a term for a sedimentary rock that has a cation (commonly Ca^{2+}, Mg^{2+}, Fe^{2+}) bonded with CO_3. Typical examples are limestone and dolomite.

Catastrophism A scientific view or frame within which natural (or in the old days supernatural) rapid and/or violent catastrophes are thought to be the major force driving observed geological changes (opening of oceans, rise of mountain ranges, evolution of species . . .). Of course, the definition of "rapid" requires ability to measure velocity, hence time and will be relative to phenomena observed by humans in their current environment.

Causal relationship A relationship involving a physically acceptable link between a cause and its effect (the effect cannot precede the cause).

Cephalopod A marine animal whose mouth and feet are in close proximity, such as the tentacled creatures octopus or ammonite.

Chaos Originally used to describe a state without ordering, the word chaos has taken a new meaning with the introduction of "chaos theory" in modern sciences. It has been found that perfectly deterministic and often rather simple causes (summarized by corresponding terms in the mathematical equations describing evolution of the phenomena) can result in apparently higher erratic (random or chaotic) behavior. The most popular image is that of a butterfly's flight in New York being able to cause drastic weather changes a few days later in Paris . . .

Chron or chronozone A span of geological time used in certain nomenclatures (biostratigraphy or magnetostratigraphy).

Coccolithophores Algae armed with delicate calcareous plates, whose fossil remains are called coccoliths.

Coelacanth A large bony marine fish that existed since the Paleozoic. Still living (although rare) in the Mozambique Straits and considered as a "living fossil."

Compensation depth A depth below which carbonates dissolve in the oceans under the effect of pressure change (at given temperature and chemical composition). Carbonate skeletons deposited below this boundary are not preserved. Under the effect of changes in temperature or acidity of ocean water compensation depth may change significantly.

Convection Transmission of energy, especially heat, through a liquid or gas by means of currents of moving molecules. When a fluid is heated enough from below, it will move in convecting rolls or cells (as in a kettle). Material in the mantle convects at large scales, with convection cells reaching at least the base of the upper mantle (at 660–670 km depth).

Core (of the Earth) The part of the Earth below 2900 km depth, consisting primarily of iron (with some nickel), alloyed to 10 to 15% in mass of several light elements (mainly silicon and sulfur, possibly oxygen, less likely hydrogen and carbon). Indeed, the density of the core is 10% lower than that of pure iron at core pressures and temperatures. The outer part of the core is molten down to 5200 km where it crystallizes into a solid inner core (the center of the Earth is at about 6400 km depth).

Crinoid A marine plant-like creature (sea-lilies) with a carbonate stem. They were abundant during the Carboniferous (Mississippian) Period.

Crust The outermost layer of the solid Earth extending down to the seismic Mohorovicic discontinuity. The oceanic crust extends down some 10 km under oceans and consists primarily of mafic ("basic") material (the top constitutes the basaltic layer generated at the mid-oceanic ridges). The continental crust is largely of granitic ("acidic") composition and extends on average to a depth of 30–40 km. It can thicken to more than 70 km depth under some mountain ranges.

D" (read "Dee double prime") The lowermost part of the Earth's mantle above the core–mantle interface, from about 2700 to 2900 km depth. Consists of mantle silicates, possibly also iron, and is thought by some scientists to be the birth place of the deepest mantle plumes. A subject of much exciting research in geophysics.

Damped oscillation An oscillatory motion of mass (like water in an oceanic wave or rock in a seismic wave), whose amplitude diminishes with time because of energy loss, in general through friction.

Daughter isotope A (parent) radioactive isotope decays into a stable (nonradioactive) isotope. The stable end-product is often called a daughter isotope.

Declination (magnetic) The angle between magnetic north (as given by the compass needle) and geographic north (in the horizontal plane, counted positive eastward).

Deus ex machina (latin) Literally "god from the machinery", by which gods were shown in the air in ancient theater. An event that comes in nick of time to solve a problem (eg an artificial explanation . . .).

Diagenesis The process in which organic or mineral material in a sediment becomes rock upon burial (water is expelled and the material hardens over geological time).

Diapir A mushroom-shaped domal structure that normally forms when a less dense liquid rises through a more dense liquid. Salt diapirs are sites where oil can preferentially be trapped and are, therefore, much sought after features in the upper sedimentary part of the crust.

Diaplectic A glassy texture, found in deformed crystals, that is produced by shock but not by melting.

Diatoms Microscopic unicellular brown algae, living in seawater or freshwater. They have an outer membrane surrounded by a siliceous shell.

Dinosaurs A group of extinct archeosaurian reptiles that thrived in the Mesozoic.

Dioxin A carcinogenic substance used in defoliants such as agent Orange.

Diplodocus (obsolete) A very large (up to 30 m) herbivorous amphibious dinosaur that lived from the Late Jurassic through the Late Cretaceous.

Dipole The simplest source of magnetic field, with a geometry resembling a bar magnet, with field lines going out at one end and back in at the other.

Dislocation An imperfection within a crystal (line defect) where slip has taken place.

Double inequality Implies that what is referred to is both respectively less than and larger than two other quantities (in the text, refers to the respective durations of an impact, human life, and volcanic eruptions in India).

Dynamo A process where the energy of motion is converted into magnetic field energy. For the Earth, fluid motion in the conducting liquid outer core generates the magnetic field we observe on the surface.

Eccentricity The amount that an ellipse departs from a circle. The Earth is in an elliptical orbit about the Sun; its eccentricity is very small and varies slowly with time under the influence of other planets.

Ecliptic The plane of the Earth's orbit, also containing the Sun. The angle of the Earth's rotational axis with respect to a pole at 90° to the ecliptic is currently about 23.5° (also equal to the latitude of the tropics). This tilt is responsible for the annual seasons.

Ejecta Material discharged by a volcano.

Extraterrestrial microspherule Very small spherical particles whose origin does not belong to the Earth.

Extrusion Used here to describe the tectonic process by which the continental mass of Indochina is laterally expelled following the collision of India with Asia.

Faraday's law The physical law (evidenced by Faraday) that a change in magnetic flux through a conducting circuit generates an electromagnetic force (potential difference).

Feldspar Silicate minerals abundant in most magmatic rocks. The calcium, sodium, or potassium content of a feldspar is used by geologists for classification and provides access to the chemical and thermodynamic conditions under which the rock was formed.

Flagellates A class of protozoans (including spermatozoids) that carry one or more flagella, or whip-like structures used in motion.

Foraminifera A microfossil that has a skeleton normally composed of calcite and bearing characteristic openings (Foramen in Latin). It lived mostly in the marine environment but could also live in freshwater.

Fourier transform A mathematical procedure to analyze periodic phenomena. Introduced by the French mathematician Joseph Fourier (1768–1830) in a study of heat propagation.

Fusulinid A Foraminifera (or Foraminiferida) that has a calcerous skeleton and is elongate like a grain of wheat. These creatures lived from the Ordovician to the Permian and were largely wiped out at the end of the Paleozoic.

Gastropods A class of mollusks with a flattened disk-shaped fleshy "foot" used in motion (e.g. snails).

Geochemistry The study of the chemical composition of Earth materials and the processes that have produced their distribution (at the level of minerals, phases, elements, or isotopes). Geochemistry has undergone tremendous progress through the application of mobilistic ideas (dynamic geochemistry), mathematical modelling, and advanced mass spectrometers.

Geochronology The dating of events in the Earth's history. Modern geochronology relies mainly on the use of couples of parent/daughter isotopes such as rubidium–strontium, potassium–argon, uranium–thorium–lead, osmium–iridium. . . .

Gondwana The southern supercontinent comprising much of present South America, Africa, India, Australia and Antarctica, resulting from breakup of Pangea.

Goniatite A group of ammonites that lived mostly from the Devonian to the Upper Permian, in which the sutures marking the edges of chambers developed lobes and saddles with a characteristic angular appearance.

Granite An intrusive igneous rock that is rich in quartz and feldspar. It commonly occurs in subduction zone settings but may be found in nearly all tectonic environments. The most abundant rock type in the continental crust.

Greenhouse effect The heating of the Earth when incoming solar radiation is radiated back from the surface (with a change in wavelength) and becomes trapped by carbon dioxide and water vapor in the lower atmosphere.

Halogen A group of highly reactive chemical elements (fluorine F, chlorine Cl, bromine Br, etc.) that combine with other elements to form volatile compounds, salts and acids. Halogen gases are emitted from volcanoes.

Heat flux The amount of heat transported per unit time through a given area.

Hiatus A break or lapse of sedimentation in a sedimentary sequence. This break constitutes a time gap as recorded by the sediments.

Hot spot A volcanic center, such as Hawaii, which most scientists believe is formed above a plume of hot mantle material that penetrates the crust. The depth of plume origin is debated. Hot spots seem to be in rather slow motion with respect to the mantle (less than 10 mm/year) and for that reason they can be useful tools for tracking plate motions.

Hydrogen chloride (HCl) A strong acid that can form naturally in volcanic environments.

Inclination (magnetic) The angle between the magnetic field direction (or the direction of magnetization in a rock) and the local horizontal plane (counted positive downwards, negative upwards). Inclination is zero at the magnetic equator and $\pm 90°$ (vertical) at the magnetic poles. A simple formula helps to derive magnetic latitudes from magnetic inclination in the case where the field is that of a dipole (which is roughly the case for the Earth). (See Chapter 3, Note 10)

Iridium (Ir) A platinum group element that is rare in the Earth's crust (a few hundredths parts per billion) but relatively abundant in iron–nickel meteorites (up to a few hundred parts per billion) and possibly in the Earth's mantle (and certainly core).

Island arc A volcanic chain of islands, like Japan or the Aleutians, that form in subduction zone environments because of melting at depth of the subducted plate, linked with high water content.

Isotope Species of the same element that have the same number of protons but different numbers of neutrons in the nucleus (e.g., ^{40}Ar and ^{39}Ar are isotopes of argon; both have 18 protons and 22 and 21 neutrons, respectively).

Joule effect The heating due to electrical current passing through a conductor. Derived from the work of English physicist James Prescott Joule (1818–89).

Kimberlite A type of rock assemblage originating in the mantle that undergoes rapid decompression during ascent to the Earth's surface. They form pipe-like structures and often contain diamonds. The name is derived from the type mines near Kimberley, South Africa.

Kinematics The physical or mathematical description of how things move.

Labradorite An iridescent blue feldspar, relatively rich in calcium, used as a decorative stone.

Lamella A thin planar intergrowth (layer) of one mineral in another.

Laterite A red soil enriched in iron or aluminium. This soil type is a chemical weathering product and forms in the tropics where precipitation is abundant.

Laurasia The northern supercontinent comprising much of present North America and Asia, resulting from the breakup of Pangea.

Limestone A sedimentary rock composed dominantly of calcium carbonate (usually in the form of calcite).

Lithosphere The rigid elastic upper 100 km of the Earth that include both the crust and the uppermost mantle. The "plates" of plate tectonics consist of lithosphere.

Lystrosaurus Fossil pig-sized plant-eater reptiles from the early Mesozoic.

Maastrichtian The final chron of the Cretaceous (and Mesozoic) between 71 and 65 Ma ago.

Magnesioferrites Minerals from the spinel family, formed from oxides of both iron and magnesium (may contain nickel).

Magnetite One of the most important magnetic minerals, with an inverse spinel structure $(Y(YX)O_4)$, in which both X and Y are iron.

Magnetometer An apparatus that measures one or several components (direction and/or intensity) of the full magnetization vector in rocks.

Mammals Warm-blooded vertebrates that breast-feed their newborn.

Mantle The silicate part of the Earth below the crust and above the outer core (2900 km depth).

Mantle plume see hot spot.

Marine regression The retreat of the sea from land.

Marsupial A mammal whose young are born unmature, then develop in a (marsupial) pouch of the adult (e.g., kangaroo).

Mass spectrometer An expensive and sensitive piece of equipment (the workhorse of any geochemistry or radiochronology laboratory) that measures the minute concentrations of individual isotopes.

Mastodon Extinct, elephant-like mammals which lived mostly in North America during the Oligocene and Pleistocene Epochs.

Mid-oceanic ridge An oceanic mountain chain formed by the upwelling of hot mantle material into a central rift valley. These are the places where new oceanic crustal material is formed and lithospheric plates drift away from each other. They often (but not always) occur in a central position in ocean basins. The total length of the world mid-oceanic ridge system is on the order of 60,000 km.

Moldavite A type of tektite found in Moldavia (western Czechoslovakia).

Mollusks Invertebrate animals with a soft body (e.g. cephalopods, gastropods).

Mosasaur A large marine lizard-like dinosaur that became extinct during the Late Cretaceous.

Nautiloid A type of cephalopod, similar to an ammonite, whose peak abundance was during the Ordovician and Silurian.

Neutron activation A technique in which chemical elements in very small amounts are identified through the spectrum of gamma rays produced by bombarding the sample with neutrons.

Neutron flux The amount of neutrons that pass through a given area. This number is important for dating schemes (such as the argon isotopes radiochronologic couple ^{40}Ar–^{39}Ar) that depend on irradiating a sample in a nuclear reactor. To obtain an age, one must know the neutron flux that passed through the sample.

Nitrogen dioxide A gas formed by combustion or lightning.

Occam's razor The principle that, given two models equally capable of accounting for observations, we should prefer the one that is most economical in terms of hypotheses or parameters.

Oceanic trench An elongate surface feature linked to subduction of one lithospheric plate under another (e.g., the Tonga, Japan, or Chile trenches). Oceanic trenches are the deepest topographic features on Earth. The Pacific ring of fire is a ring of trenches associated with subduction zones.

Oil shale A laminated brown or black sedimentary rock made of silt or clay-sized particles that contains kerogen (fossilized organic material) and from which liquid or gaseous hydrocarbons can be distilled.

Olivine A silicate rich in iron (Fe) and magnesium (Mg), with general composition $(FeMg)_2SiO_4$. This generally green mineral (sometimes of semi-precious gemstone quality) is found in basic (mafic) and ultramafic rocks (notably the oceanic crust and upper mantle).

Oort cloud The hypothetical, but very probable home of the comets, located far beyond the outermost planets. Comets episodically penetrate within the orbits of the planets, sometimes approaching Earth.

Ophiolite A section of oceanic crust and upper mantle (oceanic lithosphere) that has been emplaced through tectonic motion (generally along thrust and strike-slip faults) in a collision zone. Ophiolites are made of deep-sea sediments and basic igneous rocks altered to greenish minerals (comes from the Greek root for "snake").

Osmium (Os) A platinum group element that is rare in the Earth's crust (similar concentrations as iridium with which it is generally associated).

Ozone A blue and smelly gaseous form of oxygen. The molecule contains three atoms of oxygen. Forms in air, e.g. under the influence of an electric discharge, near transformers.

Paleomagnetism The study of the ancient Earth's magnetic field as recorded in rocks. Also the use of these fossil magnetizations to solve geological problems.

Paleontology The study of past life forms based on the fossil record.

Pangea A supercontinent comprising most of the world's continental plates that existed between about 300 and 200 Ma ago.

Paradigm "Example" in Greek. In Plato's philosophy indicates the world of ideas, considered as a model for the sensible world in which we live. In *The Structure of Scientific Revolutions* (1960), Thomas Kuhn uses the word to describe the ensemblage of postulates on which a scientific theory is founded. When too many data contradict the paradigm, a scientific revolution may occur through a shift in paradigm.

Parent isotope The radioactive isotope that eventually decays into a stable (daughter) isotope.

Paroxysmal phase A sudden and violent action (peak) of a natural phenomenon (e.g., a volcanic explosion as part of a volcanic eruption).

Perovskite One of the main mineral constituents thought to constitute the deeper regions of the mantle, likely the most abundant mineral in the Earth. Laser heating of olivine, the main constituent of the upper mantle, at high pressures in the laboratory turns it into spinel (pressures corresponding to 400 km depth) then perovskite and another mineral called magnesiowüstite (pressures corresponding to 600 km depth).

pH A logarithmic measure of the acidity of a solution, which will be neutral, acid, or basic depending on whether the pH is equal to, less than, or greater than 7.

Phosphate nodule A rounded pebble-sized nodule that forms on the sea floor and comprises mostly sand grains and marine skeletal remains, the latter rich in calcium phosphate (the mineral is commonly apatite).

Photosynthesis The cell process in which plants turn light energy from the sun into carbohydrates (sugars).

Placental mammal A mammal whose fetus develops entirely within the uterus and exchanges through a placenta (i.e., the young are born in a rather mature state). The placenta is a sponge-like blood-rich tissue attached to the uterus and communicating with the fetus through the umbilical cord.

Plankton An assemblage of very small organisms that float or drift in the upper water layers but are unable to undergo large active displacements.

Plastic soil Soil that deforms in a plastic way (more like a very viscous liquid than a solid, but with a yield strength). The soil flows but does not rupture or break.

Platform A level surface, for example a continental area covered by flat sedimentary strata overlying a basement of previously consolidated rocks.

Platinum (Pt) A rare chemical element that often occurs in nature alloyed with minor amounts of related metals (e.g., iridium, rhodium).

Plutonium (Pu) A heavy radioactive chemical element, Plutonium-244 undergoes spontaneous fission at a very fast rate. So fast that a mere 50 Ma after the creation of the solar system no more natural ^{244}Pu existed. Evidence for the isotope in meteorites, via its fission products or damage tracks, indicates that the meteorite formed soon after nucleosynthesis.

Potassium (K) The seventh most common chemical element in the crust and found in many minerals.

Precession The position of the Earth's axis with respect to a pole at 90° from the ecliptic plane is not constant. Instead, the Earth's axis moves along a conical surface around the pole to the ecliptic, under the influence of gravity of the larger planets, making one full revolution approximately every 26,000 years.

Pterodactyl A flying reptile (dinosaur) that lived from the Middle Jurassic to the Upper Cretaceous.

Punctuated equilibrium A theoretical model of evolution proposed by N. Eldredge and S. J. Gould, in which new species appear very fast after isolation of peripheral populations. They would episodically replace dominant central species that failed to evolve for long periods. This accounts for abrupt changes in the fossil record and the absence of continuous intermediate species ("missing links").

Radiolaria A class of protozoans, part of the marine plankton, with a siliceous skeleton. Thin pseudopods radiate through tiny holes; the resulting decorative pattern of the skeleton allows identification of species (both living and fossil).

Rare-earth element One of 15 elements spanning from lanthanum (La, atomic number 57) to lutetium (Lu, atomic number 71). These elements are geochemically interesting because most have the same valence (3+) yet their atomic radii decrease with increasing atomic number. The ratios of these elements in rocks and minerals can yield information on their sources.

Refractive index The ratio of the velocity of light in a medium (e.g., a crystal) to that within a vacuum. This property helps petrologists to distinguish different minerals in thin section under a microscope. The refractive index of water is larger than one, giving the illusion that a stick is kinked at the air–water interface.

Rhenium (Re) One of the most refractory and densest white metal elements, a member of the platinum group elements.

Rhyolite A light-coloured volcanic rock consisting mainly of alkali feldspar, and free silica, with minor amounts of mafic minerals; the extrusive equivalent of granite.

Richter scale A logarithmic scale that expresses the magnitude of an earthquake by relating the measured amplitude of ground deformation to the distance of the seismometer from the earthquake source. The magnitude characterizes the total energy released by the earthquake.

Ruthenium (Ru) A hard solid metal melting only at high temperature, a member of the platinum group elements.

Saurian Reptiles with scales. Lizards and iguanas are living examples, plesiosaurs are extinct saurians.

Sauropods A group of fully quadripedal, seemingly herbivorous dinosaurs from the Jurassic and Cretaceous (small heads, spoon-shaped teeth, long necks and tails, columnar legs).

Scaglia rossa Literally "pink scale" in Italian. A sequence of red, homogeneous limestone and (flaky) marly limestone of Turonian (upper Cretaceous) to Eocene age exposed in the Umbrian Appenines of Italy.

Secondary convection Smaller-scale, or secondary convection can occur, for instance, below older oceanic lithosphere, forming elongate rolls that can be traced in the long wavelength undulations of sea-floor topography or gravity field.

Secular change (magnetic) The slow changes in the Earth's magnetic field recorded on the time-scales of decades to centuries. For instance, because of secular variation, declination in Paris will have changed from $22°$ E in 1800 to about $2°$ E in the year 2000 (i.e. $1°$ per 10 years on average).

Seiche Damped oscillations triggered by an earthquake in a (semi-) closed oceanic basin, similar to those that develop when one suddenly agitates a big basin full of water and then lets the water return to equilibrium.

Seismology Both the science of earthquakes and the study of seismic wave propagation in the Earth.

Selenium (Se) A non-metallic chemical element that substitutes readily for sulfur (S). It is common in native sulfur deposits of volcanic origin and is also present in seawater, sedimentary rocks, and meteorites.

Shatter cone A striated conical rock fragment along which fracturing has occurred. Diagnostic of passage of a shock wave through the rock, usually because of meteorite impact.

Shocked mineral A mineral that has been modified by passage of a shock wave through it. This generates characteristic features (dislocation planes, glass lamellae, crystal twinning) seen only under more or less powerful magnification.

Signal processing A set of mathematical techniques to extract information from or modify (filter) a signal. The signal usually consists of a series of numbers representing evolution in time (or space) of some physical or chemical parameter (e.g., amplitude of a sound wave, electrical potential, changes in number of species, numbers of craters, evolution of temperature, gas composition, climate, etc.).

Signor–Lipps effect Backward smearing of extinction records owing to imperfect sampling of the fossil record. Rarer fossil species are less likely to have been found close to their actual upper level of existence and, therefore, appear to have become extinct earlier. This makes a catastrophic extinction appear to be gradual. Introduced in 1982 by paleontologists P. W. Signor and J. H. Lipps.

Silica Chemical composition SiO_2. Silica is one of the most common and strongest building blocks of many minerals and is found itself in naturally occurring forms such as quartz, chert, agate, etc.

Smectite A greenish type of clay mineral.

Spinel Spinel refers to both a mineral $(MgAl_2O_4)$ and a mineral structure (XY_2O_4). The mineral is an accessory product in mafic rocks and certain types of metamorphic rock. Magnetite, one of the most important magnetic minerals, has an inverse spinel structure. Olivine turns to spinel structure between 400 and 600 km depth.

Stegocephalian A striking large dinosaur; a popular but obsolete term that has been replaced by Labyrinthodonta.

Stishovite A high pressure form of quartz (SiO_2) commonly found in meteorite impact structures.

Stratigraphy The study of formation, composition, sequence, and correlation of stratified rock layers.

Stratosphere The outer layer of the atmosphere that overlies the trophosphere (above 11 to 18 km depending on latitude and season).

Strontium (Sr) A silver-white alkaline earth metal element that is soft like lead.

Subduction trench See oceanic trench. The term subduction refers to passage of one lithospheric (usually oceanic) plate beneath another owing to convergence of the two plates.

Sulfide A metallic or semi-metallic sulfur-bearing salt such as pyrite (fool's gold, FeS_2)

Sulfur dioxide (SO_2) A toxic colorless but irritating gas formed, for instance, by volcanic activity or organic decay.

Sulfuric acid (H_2SO_4) An acid found in volcanic environments, resulting from oxidation and hydration of sulfur dioxide. It can also form from reactions with factory emissions (e.g., coal contains varying amounts of sulfur).

Superchron A long period with no change in magnetic polarity. The Earth's magnetic field has reversed on average four times every 1 Ma since 5 Ma ago. However, the reversal rate has not been constant, far from it. Paleomagnetists have found two long periods, or superchrons, when magnetic polarity stayed constant for tens of millions of years. One was called the Kiaman Reversed Superchron (the field at that time stayed reversed) and lasted for about 70 Ma from the mid-Carboniferous to the end of the Permian. The other was during the Cretaceous, from about 118 to 83 Ma ago. That 35 Ma interval of normal polarity is referred to as the Cretaceous Long Normal Superchron.

Supernova An explosion that occurs as the interior of a massive star collapses under its own gravity and the upper layers are blown away.

Tectonic plate Any one of the internally rigid (elastic) blocks of lithosphere (crustal and mantle parts) that move horizontally across the Earth's surface relative to one another.

Tektite A form of dark (melted) silicate glass of nonvolcanic origin linked to meteorite impact.

Terrella The name given in the Renaissance (e.g., by William Gilbert in England around 1600) to a small sphere carved from magnetite used to study magnetic properties of materials and infer properties of the whole Earth (literally "small earth" in Latin).

Tethys (Sea or Ocean) The ocean that existed between the Eurasian and African, Arabian, and Indian plates. It vanished when they collided, generating the Alpine and Himalayan mountain ranges. The Mediterranean is a remnant of the Tethys, currently undergoing closure at about 10 mm/year.

Thorium (Th) A dense, radioactive gray metal that decays to lead.

Tomography A technique in seismology used to generate three-dimensional pictures of the Earth's interior (made from measuring various seismic waves at different receivers, or seismographic stations). What is portrayed is velocity of pressure waves or shear waves through material or elastic properties such as rigidity. The procedure is similar to CAT scanning in medicine and was actually invented before (although the latter, based on it, has become far better known to the general public).

Transform fault A type of strike–slip fault that forms a boundary between two tectonic plates as one slides horizontally past the other. These faults link, or "transform", other types of plate boundary (mid-oceanic ridges and subduction zones) one into the other.

Transition zone A zone in the mantle lying approximately from 400 to 660–670 km depth. Seismic wave velocity increases in this region, indicating a change in mineralogy to denser mineral phases.

Traps A geographical term to indicate huge thick and flat expanses of lava flows, usually bordered by step-like escarpments owing to erosion. The root of the word appears in several Nordic languages, with the sense of "stair steps." The term was introduced by Bergman, a Swede, in 1746.

Triceratops A very "popular" species of three-horned dinosaur that became extinct in the Late Cretaceous.

Trilobite A marine animal (arthropod) resembling a large insect with a hard dorsal skeleton (divided both laterally and longitudinally into three segmented regions) that was easily preserved. These creatures lived during the Paleozoic (mostly during the Cambrian and Ordovician) and became extinct at the end of the Permian.

Tropopause The upper limit of the troposphere marked by an abrupt minimum in temperature.

Troposphere The part of the atmosphere from ground level to about 11 to 18 km altitude (depending on latitude and season).

Tsunami Tidal waves triggered by earthquakes (from the Japanese).

Turbidity current A density current at the bottom of the ocean resembling an underwater avalanche. It occurs when sediment lying on a steep slope begins flowing downhill. The sediment is dispersed to lower, more gently sloping parts of the ocean floor, where it undergoes diagenesis and produces a characteristic sedimentary rock called a turbidite.

Tyrannosaurus rex One of the most famous of the large carnivorous dinosaurs that lived in North America and Asia before becoming extinct at the end of the Mesozoic.

Uniformitarianism A scientific view or frame within which observed natural (e.g., geological) phenomena are thought to have resulted from slow geological changes (with velocities typical of those observed today). The large amplitude of geological effects would then result from accumulation of large slices of geological time. One further distinguishes substantive uniformitarianism, in which it is assumed that the amplitudes of geological phenomena observed by humans represent the whole range of permissible amplitudes over geological time (for instance, volcanic flows and eruptions seen in present-day volcanoes would be typical of all flows and eruptions in the geological past; observation of traps shows that such cannot be the case).

Upper mantle The part of the mantle below the crust and above 600–670 km.

Volatile Elements that vaporize at relatively low temperatures. Volatile elements can also reduce melting temperatures.

Williams–Riley A form of biological pump. Under normal conditions, single-cell algae extract carbon dioxide from the air and water and use it to build their calcareous skeletons.

When they die, their solid remains fall to the bottom and are incorporated ("pumped") into limestone sediments.

Zinc (Zn) A chemical element, used mostly in metal alloys, that commonly precipitates where crustal fluids pass through carbonate rocks.

Zircon A zirconium silicate ($ZrSiO_4$) that forms as an accessory mineral in igneous and metamorphic rocks. Uranium substitutes easily for zirconium in the zircon structure, but lead (Pb) does not. This relationship allows Earth scientists to perform age dating on single zircon crystals because there is no (or very little) initial daughter isotope (Pb) concentration present to influence the measured date.

Index of Authors

Subject index